Dychweler y llyfr hwn erbyn y dyddiad isod,
neu pan elwir amdano gan y Llyfrgellydd.

NITROGEN AND CARBON METABOLISM

DEVELOPMENTS IN PLANT AND SOIL SCIENCES

VOLUME 3

Other titles in this series:

Volume 1
Soil-Water and Nitrogen in Mediterranean-type Environments
edited by J. Monteith and C. Webb
ISBN 90-247-2406-6

Volume 2
Nitrogen Losses and Surface Run-off from Landspreading of Manures. Proceedings of a Workshop in the EEC Programme of Coordination of Research on Effluents from Livestock, held at The Agricultural Institute, Johnstown Castle Research Centre, Wexford, Ireland, May 20-22, 1980
edited by J.C. Brogan
ISBN 90-247-2471-6

In preparation:

Structure and Function of Plant Roots. Proceedings of the Second International Symposium, held in Bratislava, Czechoslovakia, September 1-5, 1980
edited by R. Brouwer, O. Gasparikovấ, J. Kolek and B.C. Loughman

Series ISBN 90-247-2405-8

NITROGEN AND CARBON METABOLISM

Proceedings of a Symposium on the Physiology and Biochemistry of Plant Productivity, held in Calgary, Canada, July 14-17, 1980

edited by

J. DEREK BEWLEY

Department of Biology,
University of Calgary,
Alberta, Canada

1981

MARTINUS NIJHOFF / DR.W. JUNK PUBLISHERS
THE HAGUE / BOSTON / LONDON

Distributors:

for the United States and Canada
Kluwer Boston, Inc.
190 Old Derby Street
Hingham, MA 02043
USA

for all other countries
Kluwer Academic Publishers Group
Distribution Center
P.O.Box 322
3300 AH Dordrecht
The Netherlands

This volume is listed in the Library of Congress Cataloging in Publication Data

ISBN 90-247-2472-4 (this volume)
ISBN 90-247-2405-8 (series)

PRINTED IN THE NETHERLANDS

CONTENTS

Preface VII

Chapter 1.
Profound and Useful Discoveries.
R.G.S. BIDWELL 1

Chapter 2.
Comparison of the Efficiency of Nitrate and Nitrogen Fixation in Crop Yield.
J. RIGAUD 17

Chapter 3.
Nitrate Uptake, Production and Transport in the Whole Plant.
L.E. SCHRADER and R.J. THOMAS 49

Chapter 4.
Carbon and Nitrogen Partitioning in the Whole Plant - a Thesis based on Empirical Modeling.
J.S. PATE and D.B. LAYZELL 94

Chapter 5.
Relationships between Nitrogen Metabolism and Photosynthesis.
J.A. BASSHAM, P.O. LARSEN, A.L. LAWYER and K.L. CORNWELL 135

Chapter 6.
Photosynthate Limitation of Symbiotic N_2 Fixation.
K. ANDERSEN, S.T. LIM, H. SPILLER, K.T. SHANMUGAM and R.C. VALENTINE 164

Chapter 7.
Photorespiration and Nitrogen Metabolism.
D.T. CANVIN 178

Chapter 8.
Seed Storage Proteins: Genetics, Synthesis, Accumulation and Protein Quality.
B.J. MIFLIN and P.R. SHEWRY 195

PREFACE

This book results from a symposium on the theme of 'The Physiology and Biochemistry of Plant Productivity' which was held at the University of Calgary from July 14-18, 1980, and was jointly sponsored by the Canadian Society of Plant Physiologists and the International Association of Plant Physiologists.

The subject matter of the book deals with various aspects of nitrogen and carbon metabolism, their interrelationships and interdependence. The topics covered in the chapters highlight various interesting and important lines of research that are in progress. There is no attempt to provide a comprehensive coverage of the basic physiological knowledge upon which this research depends - important references are to be found at the end of each chapter, however, and the reader will be able to pursue these as necessary.

An introductory chapter by Dr. R.G.S. Bidwell (winner of the C.S.P.P. Gold Medal in 1979) considers some implications of plant physiological research and the aims and responsibilities of plant physiologists. In the next two chapters Drs. J. Rigaud and L.E. Schrader (with R.J. Thomas) elaborate on current research on nitrate metabolism and nitrogen fixation, and how an understanding of these phenomena might be usefully applied towards the manipulation of plants to improve productivity. Dr. J.S. Pate (with D.B. Layzell) introduces the topic of carbon and nitrogen interrelationships and develops models for their partitioning and transport within plants. Continuing this theme Drs. J.A. Bassham (with P.O. Larsen, A.L. Lawyer and K.L. Cornwell) and R.C. Valentine (with K. Andersen, S.T. Lim, H. Spiller, and K.T. Shanmugam) discuss the convergent pathways of carbon and nitrogen metabolism, the regulation of carbon flow from photosynthesis into sugars and amino acids, and how photosynthate production might limit symbiotic nitrogen fixation, an event which has a high metabolic energy cost. Photorespiration is also closely and importantly associated with nitrogen metabolism and Dr. D.T. Canvin explains the essential links between these processes. In the last chapter, Dr. B.J. Miflin (with P.R. Shewry) details the nature of seed storage proteins, a vital dietary source of amino acids, their synthesis and genome organization, information which is necessary to accurately assess the prospects for improving protein quality.

These chapters are based on papers presented at two symposia at the CSPP/IAPP conference, and I am grateful to Drs. R.H. Hageman and L. Beevers for their role as Chairmen of these sessions. The programme was arranged with the excellent cooperation of Drs. A. Oaks and F. Wightman, and the meeting itself relied upon the organizational skills of Dr. R.P. Pharis. Cooperation by Dr. A. Quispel on behalf of the IAPP is appreciated. I am very grateful to the authors for the prompt manner in which they provided me with their manuscripts, and to Erin Smith of the Biology Dept., University of Calgary for her secretarial skills and help in the final preparation of this book.

September 1980

J. Derek Bewley
Professor of Biology
University of Calgary
Calgary T2N 1N4
Alberta, Canada

CHAPTER 1

PROFOUND AND USEFUL DISCOVERIES

R.G.S. Bidwell
Wallace RR #1, Nova Scotia, Canada BOK 1YO

Introduction
Tank Cultivation of Marine Algae: Background
Gas Exchange in Marine Algae: The Air Suspension System
Seaweed Cultivation: The Transfer of Ideas
Industrial Research: Funding, Operation and Control
References

J.D. Bewley (ed.), Nitrogen and Carbon Metabolism

The Gold Medal Address was presented at the meeting by Dr. R. G. S. Bidwell, recipient of the Canadian Society of Plant Physiologists Gold Medal (1979) in recognition of his outstanding published contributions and for his service to plant physiology and the Canadian Society.

INTRODUCTION

The title of this chapter is not intended to refer to my own work, but is taken from an observation by J. B. Passioura, C.S.I.R.O., Canberra (11):

> Plant physiologists have two responsibilities to the public whose money supports them. One is to make profound discoveries. The other is to make useful ones.

A discussion of the physiology and biochemistry of plant productivity implies practical application of laboratory research. There have been a number of notable successes in the transfer of laboratory research to agricultural

[1] I gratefully acknowledge continued research support from the National Research Council of Canada while I was on the staff of Queen's University, Kingston, Ont. I am also deeply grateful to the Director of the N.R.C.C.'s Atlantic Research Laboratory, Halifax, N.S., for my appointment as Visiting Scientist during a Sabbatical year in 1976-7, and as Guest Scientist at intervals since then.

technology. But there have been a great many more failures. To put the matter in a different perspective, we have made many profound discoveries that are inherently useless or inaccessible for the development of agricultural technology. This is not because they are bad science, or improperly done, or even in remote or inaccessible fields. It means that while most of us are more or less adept at making profound discoveries, few of us are good at making useful ones.

I do not wish to belittle the importance of pure science. That activity leads, or should lead, to the profound discoveries that are essential for the continuation and development of science. Profound discoveries are able to take care of themselves because they generate new ideas, new questions, and new research. There are no dead ends in pure science. But they are not necessarily useful.

Profound discoveries have another attribute: important ones are not made very often, or by very many scientists. This leaves the great majority of us with the sensible option of making useful discoveries. But considering the large amount of work done and the vast sums of money that support it, the size of the return in useful discoveries seems to be remarkably small. This suggests that perhaps many of us are not familiar with the techniques of making useful discoveries. It is my purpose in this paper to describe, as a case history, some research that was started with the hopeful intention of making profound discoveries but which, through a process that was largely serendipitous, began to make useful discoveries instead, and ultimately lead to a genuine (if small) break-through in agricultural technology. I refer to the technology of cost-effective on-shore cultivation of marine algae. My interest here is not so much in the research itself as in our discovery (hopefully profound) of how to make useful discoveries.

TANK CULTIVATION OF MARINE ALGAE: BACKGROUND

The harvesting of marine algae has always been an important but small industry on the east coast of Canada. The most important product is carrageenan, largely derived from the red alga Chondrus crispus. Much research has been done on seaweed at the Atlantic Research Laboratory of the National Research Council of Canada, giving this institute a well deserved world-wide reputation in this field. But few of the discoveries from this research have been directly useful to industry.

Some years ago Dr. A. C. Neish began to develop the concept of farming seaweed at A.R.L. His pioneer work led to the first real breakthrough: the discovery that seaweed could be cultivated in tanks, and that a high-productivity, non-sporulating variety could be selected and maintained, apparently as a clone (10). A patent was taken out on this process. Dr. Neish was awarded the first Gold Medal of the Canadian Society of Plant Physiologists in 1970, and his Gold Medal Address, given in Toronto in 1971, was about seaweed cultivation. It is a great honor for me to be able to follow Dr. Neish in every respect but one: his discoveries were profound, mine are at best useful.

In spite of the N.R.C.'s patent and the vision of Neish's work, the system was not easily accessible to industry. Capital costs for tanks were very large. Energy requirements for stirring the tanks and preserving a suitable environment for growth were enormous. And seaweed growth was inadequate; certainly far slower than the plants are capable of in a natural environment. Neish's contribution was to show that the thing was possible. But something more was needed to produce seaweed that did not cost ten or twenty times more than the still adequate wild harvest.

Three main lines of research led ultimately to the situation where an effective industrial technology for onshore seaweed production could emerge. The first was a careful analysis and development of the engineering technology for stirring or suspending the plants, conducted by the Genu Products Canada Ltd. research team at Halifax, N.S. This led ultimately away from the cumbersome, energy-inefficient, and biologically unsuitable mechanical stirring developed at A.R.L. to a modular air-suspension system that not only optimizes cost and the biological conditions for maximal yield, but also permits ready control of carbon fertilization. The second was a massive research program to determine the optimal grid of all major controllable environmental factors in outdoor tanks (nutrients, temperature, flushing rates, air-sparging rates, etc.), also conducted by Genu. The third developed from research on photosynthesis and photorespiration of seaweeds done at A.R.L. while I was on sabbatical leave and a guest in the laboratory of Dr. J. L. McLachlan (4).

All three research programs were going on simultaneously. However, it was the accidental interaction between industrial and academic research that led to the coordination of the third program with the first two. This in turn permitted a one to two order of magnitude increase in the speed of development and the rate of data acquisition, and provided the answer to the key problem of why the plants would not grow as rapidly in culture as they could in the sea. Other programs were also important, such as the development of selective herbicides to eliminate epiphyte problems, the selection of varieties and strains to improve growth and carrageenan production, and the development of rapid, simple analytical procedures for carrageenans so that systems could be optimized for the production of commercially desirable carbohydrates, and not merely biomass (3). However, all these depended on the growth and production-analysis technology that developed from our cooperative research (7).

I must stress that I make no claim for unique success in our academic-industrial cooperation. This is intended to be a record of my own discovery of the way that such cooperation can be made to work successfully. I am aware of many other successful ventures. But, the contrast between the large amount of excellent scientific research now being done in this country and the relative paucity of successful scientific-agricultural-industrial cooperation leads me to believe that an account of the factors required for success, and those that lead to failure (or to an endless continuation of excellent but useless science, which is the same thing), might be of value. Doubtless there are other ways of doing it. This is the way we did it.

GAS EXCHANGE IN MARINE ALGAE: THE AIR SUSPENSION SYSTEM

Some years ago Dr. W. A. Dodd, then a post-doctoral fellow in my laboratory, and I needed to do $^{12}CO_2$-$^{14}CO_2$ pulse-chase experiements with chloroplasts. This is not possible in an aqueous suspension medium, so we developed the technique of suspending chloroplasts in air on a non-absorbant mesh which we called the artifical leaf (5). The technique was extended by us and others to measure gas exchange in unicellular algae (1,6). I wished to examine the gas exchange, particularly photorespiration, in marine algae. Earlier work showed that these organisms have more or less conventional glycolate metabolism (13), and some measurements made under rather less than satisfactory conditions suggested that they photorespire (14).

It is, however, very difficult to measure gas exchange in seaweeds with accuracy because the sea is a complex medium containing, among other things, 2 mM carbonates. This causes two immediate problems: even a slight pH change causes uptake or liberation of carbon dioxide by the medium, which affects photosynthesis measurements. It also causes a change in the $CO_2:HCO_3^-:CO_3^=$ equilibrium in the medium, as well as quite substantial changes in the concentration of each molecular or ionic species, which may affect photosynthesis itself. In addition, there are two further problems that are not always recognized. Marine algae are apparently able to assimilate either carbon dioxide or bicarbonate (or both). No data are yet available that show if the assimilation rates are the same, or if seaweeds can actively assimilate bicarbonate like some freshwater algae (8,12). Thus any factors, such as pH changes or even the process of photosynthesis itself, that may affect the $CO_2:HCO_3^-:CO_3^=$ ratio may also affect photosynthesis. Furthermore, the plants are absorbing carbon dioxide or bicarbonate from a solution that is being replenished by the addition of carbon dioxide from sparged air. Regardless of the rate of solution of carbon dioxide into water, which is actually quite slow, or the vigor and fineness of air sparging, the rate of carbon dioxide replenishment must be less than the maximum rate of photosynthesis. Even under ideal conditions, the steady-state concentration of carbon dioxide or bicarbonate in the medium can reflect only a dynamic equilibrium that depends on the mass of plants, their photosynthetic capacity, the light intensity, temperature, and all other factors that might affect photosynthesis, as well as the volume of water and the efficiency of sparging. Since photosynthesis depends on the concentrations of carbon dioxide and bicarbonate, the rate of photosynthesis will depend upon all of these factors. Measurement of photosynthesis of plants in seawater thus frequently gives only the rate of carbon dioxide absorption by the water.

These problems can all be circumvented by using the technique of the artificial leaf and measuring gas exchange of the plants while they are suspended in flowing, water-saturated air. By doing this, we learned certain things about gas exchange of seaweeds: they do not normally photorespire (although some have a strong, oxygen-sensitive wound respiration that mimics photorespiration), their rates of gas exchange and their biochemical processes are identical in air or water suspension, and they are capable of much higher rates of photosynthesis than had hitherto been suspected. However, these are not of concern here, and are in preparation for publication elsewhere (see 4).

What we did recognize that is of critical importance is that most of the seaweeds that we examined rely primarily on the bicarbonate in the sea for their carbon nutrition, although they can absorb carbon dioxide. This is not surprising, given that the equilibrium concentration of bicarbonate (approximately 2 mM) exceeds that of carbon dioxide by a factor of 200 in the sea. Also, being ionic, bicarbonate ions are susceptible of active absorption by ion-pumping mechanisms, which carbon dioxide is not. This may possibly provide an explanation for the fact that seaweeds do not photorespire (2).

A consequence of their reliance on bicarbonate in the sea is that air-suspended seaweeds need a very high level of carbon dioxide for normal rates of photosynthesis. Normal rates (which can be maintained in the sea for long periods because of the large volume and movement of that medium) can be measured in seawater-suspended plants by artificially maintaining the pH and bicarbonate content. Then the concentration of carbon dioxide in air required to give the same rate can be determined experimentally. It is usually in the range of 1500 to 2000 $\mu l\ l^{-1}$. Photosynthesis can be measured under these conditions quickly and repetitively, and rates of gas exchange so measured are sufficiently close to the rates maintained under natural conditions that they can be used to predict growth over a period of several weeks in plants cultured under natural conditions. Predictions are usually accurate to within approximately 10%, and measurements can be made in a few minutes (7). The convenience, speed, and accuracy of this technique, as well as the underlying principle of potential limitation in carbon nutrition, provided the basis for the transfer of laboratory research to industrial technology.

The importance of the problem of carbon limitation should be stressed. Aside from the possibility that many other nutritional factors may well have been investigated under conditions of limited carbon nutrition (9), accurate estimation of productivity is impossible unless this problem is recognized. The following measurements indicate how great it may become. Seawater normally contains slightly over 2 mM total carbonates. In a typical laboratory culture situation, 2.0 g fresh seaweed (*Gracilaria tikvahiae*) in 100 ml seawater medium was aerated by a flowing air stream injected into the water at a rate which gave sufficient bubbling to stir, agitate and suspend the seaweed. After three hours of illumination at *ca* 50 $\mu E\ m^{-2}\ s^{-1}$ (perhaps 5% of the equivalent of full sunlight) the concentration of total carbonates in the medium had fallen to 0.15 mM, at which concentration photosynthesis was only about 10% of the normal rate in fresh seawater. The point is that the capacity of the plant to absorb bicarbonate was very much greater than the capacity of the seawater to absorb carbon dioxide from the air. The dynamic equilibrium that was eventually established left the plant grossly undernourished with respect to carbon. However, this problem escaped detection until these measurements were made because carbon malnutrition does not produce any obvious symptoms except slow growth.

SEAWEED CULTIVATION: THE TRANSFER OF IDEAS

The account to this point seems straightforward and uncomplicated because it is told with the advantage of hindsight. The actual situation was quite different. In my ivory tower at the university and as a Visiting Scientist at the A.R.L. I was unaware of any details of the seaweed cultivation industry, its research programs, or its problems. Equally, they had never heard of me or my research program on algal photosynthesis and photorespiration. If ever there were two solitudes, they are this country's research scientists and its industrialists.

The meeting of these two solitudes occurred at a social function when I was introduced to the General Manager of Genu Products Canada Ltd., and subsequently invited to visit his research station. Two things were immediately evident. The first was that the research program, though carefully planned and imaginatively executed, could not provide reliable data because experiments took several weeks to perform and the experimental conditions, being the natural outdoor environment, were anything but constant. The second was that the tank-cultivated plants, like the cultured Gracilaria mentioned above, were suffering from carbon malnutrition.

The question of carbon nutrition could be, and was, tested immediately with gratifying results: growth rates could be improved by 25 to 50% by carbon fertilization. Ultimately a system of automatic, self-demand carbon fertilization has been devised that is not only effective in making the plants grow at maximum rates, but is also cost effective - a requirement that is often overlooked.

The second problem, that of developing a system that could rapidly and accurately predict growth and productivity, took somewhat longer. However, with the help of Dr. N. D. H. Lloyd, who had taken charge of the biological program, we developed a system that permitted rapid, complex, multivariate analysis of all controllable environmental parameters at the growth-cabinet level. This could then be scaled up to the outdoor tank system in order to optimize cost-effective production. The development of simple, rapid techniques for the quantitative and qualitative analysis of carrageenan, the desired product, made possible the cost-optimized production of carrageenan, rather than algae. It turns out, as might be expected, that the two are by no means the same.

A further development was the institution of "brainstorming" sessions consisting of the biological and technological leaders of the company team, a plant physiologist consultant, a member of the Marine Botany group from A.R.L., a consulting engineer, and the company's market analyst, all under the direction and leadership of the company's general manager. This arrangement permits the best possible use of the company's research resources to optimize the system, and provides continuous cost-benefit analysis of all potential innovations as well as quality control of both research and production management. It has two further consequences: it keeps the company's research program under tight rein and strongly goal oriented, and it permits continuous integration and optimization of the technological, engineering, biological, and production aspects of the system. Furthermore, interaction of the various disciplines often produces valuable new ideas that might never have been derived from a single viewpoint.

INDUSTRIAL RESEARCH: FUNDING, OPERATION AND CONTROL

Once again, the arrangement sounds quite simple when viewed with hindsight. But it was developed by trial and error, and it seems probable that many of the errors might be avoided in future. Also, once we had worked out the right way of going about it, the reasons for the many previous failures became obvious. I hope that the short outline provided below will help others from repeating our mistakes.

The two major problems that need to be faced, when one wishes to develop an interactive research program involving research science and a small industry or agricultural-technology organization, are how it should be paid for and how to get it started. The major expenses are facilities, equipment, and personnel. Doing research is very costly, and small organizations do not have the necessary large resources. However, there are ways of getting around this difficulty.

Capital costs need not be a problem. Laboratories are available, in universities or government institutions, where most kinds of research can be done. Usually major equipment can be found that will do needed analyses or experiments either free of for a nominal charge.

Research personnel at the scientific level presents a more difficult problem. The experience, capability, and synthetic ability of senior scientists is needed, but is too expensive for most small companies to hire fulltime. Contracting out research to university or government scientists is seldom effective, because the ivory-tower scientist seldom appreciates the specific problems of industry: priorities, timing, cost-benefit analysis, etc. Government research laboratories, in spite of their acknowledged role in helping industry, are all too frequently subject to the same problems. It has usually been found counterproductive for small companies to hire young doctoral graduates. They are frequently inexperienced, not easily able to accept the rather rigid yoke of industrial research, and often primarily concerned about an academic future, and hence about publishing papers.

The answer is consultation. We have had great success with think tanks consisting, for example, of an engineer, a technologist, a biologist, and a market analyst, under the leadership of an industrialist. The brainstorn effect from the interaction of such a group results in high output for relatively small cost.

Research programs developed by such a group, and periodically refined as required by new input, can be conducted in an appropriate government or university laboratory by technical personnel. Much of the costs, both of the consultants and of the scientific staff, can usually be covered by government research and development assistance programs.

The key point in this scheme is that the research program must be under the direction of the industrial leadership. Industrialists understand the limitations imposed by industry. They, ultimately, stand or fall by the results of the research. Therefore, they must have the responsibility of deciding research priorities. It is the job of the scientist to establish the validity of proposed research, and of the engineering and technology personnel to establish its value and feasibility. But the overall direction must lie with the industrialist. In this way useful, cost-effective research gets done first. Lines of research, however interesting, that are not leading to useful developments can be dropped or shelved.

The second, and I think more difficult, question is how to start or organize a potentially useful interaction. In spite of attempts by the government, the N.R.C.C., and other granting bodies to increase the proportion of goal-oriented research done by Canadian scientists, the overall level of success has been disappointing. There are three obvious ways by which this situation might be improved.

The first is that scientists should seek out industrial or agricultural businesses doing things that potentially relate to their field of expertise. Nobody knows as well as the individual scientist what he is capable of. The industrialist will never know in what ways a scientist could be useful or valuable to him unless the scientist tells him.

The reciprocal should, of course, be true: industrialists ought to comb university and government laboratories in related fields to find scientists who might help them. To that end, it would be very useful if university departments, in particular, would advertise (as they do for graduate students) the scientific expertise available on their staff that might prove industrially useful. Industry and agriculture need this encouragement to help them break down the doors of academe. In addition, it is to be hoped that the new orientation of government research, particularly of the N.R.C.C., towards industrial goals will result in better liaison between industrialists and scientists. We need lists of identifiable problems in want of solution, and lists of problem solvers.

Finally, there is need for revision of the system of public recognition and reward to scientists. The major quantitative index of academic success is still publication. Unfortunately, much industrially useful research effort does not lead to publications, or may even be suppressed for reasons of confidentiality. This tends to discourage young research scientists from entering an industrial career, and inhibits many from whole-hearted pursuit of industrial research goals. Money does not seem to be a useful substitute, and in fact the salaries paid to junior industrial research workers are frequently not competitive. It is true that the Royal Society of Canada now has an Applied Sciences section in its Academy of Sciences, but this represents only a small impetus applied at the wrong end of the scale. Perhaps the solution is to convince leaders of industry and technology to give public awards to scientists who have contributed to the solution of specific problems or to the advancement of technology. This would require some pressure and coordination, but if it were seen to be a productive or, in a real sense, a cost-effective effort, doubtless some system of awards could be established.

This consideration applies equally to government scientists. They are judged by their peers, and frequently rewarded, by the same standards applied to university scientists. This powerfully constrains their willingness to do unpublishable industrially oriented research, and makes it necessary for them to do research that will lead to publications in order to maintain and develop their scientific reputation. The result is counterproductive from the industrial point of view, because industry needs immediate solutions for specific problems and cannot wait for the completion of a publishable research program. Clearly, the system for assessing and rewarding scientific merit, particularly for those scientists who wish to or must do applied research, is badly in need of revision.

Ultimately, however, it is the responsibility of the research scientist to demonstrate what he can do, and to find the industrial concern that needs him to do it. To return to the major point: one must not belittle or abandon pure science. That is still essential to make the profound discoveries by which science and technology progress. But it is an obligation of scientists who are supported by the public to make useful discoveries, as well as profound ones.

REFERENCES

1. Bidwell, RGS, Photosynthesis and light and dark respiration in marine algae. Can J Bot 55: 809-818, 1977
2. Bidwell, RGS, The carbon dioxide machine, or plants on the make. NS Inst Sci 28: 26-34, 1977
3. Bidwell, RGS, Toward improved understanding of polysaccharide synthesis and storage in marine algae. In: Synthetic and Degradative Processes in Marine Macrophytes, Srivastava, LM, (ed.), in press

4. Bidwell, RGS, JL McLachlan, and NDH Lloyd, Rapid infra-red analysis of carbon dioxide exchange in marine algae. Proc Can Soc Plant Physiol 20, 1980.

5. Dodd, WA, and RGS Bidwell, Photosynthesis and gas exchange of Acetabularia chloroplasts in an "artificial leaf". Nature 243: 45-47, 1971

6. Lloyd, NDH, DT Canvin, and DA Culver, Photosynthesis and photorespiration in algae. Plant Physiol 59: 936-940, 1977

7. Lloyd, NDH, JL McLachlan, and RGS Bidwell, A rapid infra-red carbon dioxide analysis screening technique for predicting growth and productivity of marine algae. Proc Xth Internat Seaweed Symp, Göteborg, in press

8. Lucas, WJ, RM Spanswick, and J Dainty, HCO_3^- influx across the plasmalemma of Chara corallina. Plant Physiol 61: 487-493, 1978

9. McLachlan, JL, Inorganic nutrition of marine microalgae in culture. In: Synthetic and Degradative Processes in Marine Macrophytes, Srivastava, LM (ed.), in press

10. Neish, AC, EF Shacklock, CH Fox, and FJ Simpson, The cultivation of Chondrus crispus. Factors affecting growth under greenhouse conditions. Can J Bot 55: 2263-2271

11. Passioura, JB, Accountability, philosophy and plant physiology. Search 10: 347-350, 1979

12. Radmer, R and O Ollinger, Light-driven uptake of oxygen, carbon dioxide, and bicarbonate by the green alga Scenedesmus. Plant Physiol 65: 723-729, 1980

13. Tolbert, NE, Photorespiration, In: Algal Physiology and Biochemistry, Stewart, WPD (ed.), University of California Press, Berkeley, 1974

14. Tolbert, NE, and CB Osmond, (eds.), Photorespiration in Marine Plants, University Park Press, Baltimore, Md, 1976

CHAPTER 2

COMPARISON OF THE EFFICIENCY OF NITRATE AND NITROGEN FIXATION IN CROP YIELD

J. Rigaud
Laboratoire de Biologie végétale, Faculté des Sciences
et des Techniques, 06034 NICE Cedex, FRANCE

Introduction
Nitrate utilization by legumes
Nitrate absorption
The sites of nitrate utilization
Nitrate reduction in nodulated roots
Nitrate reduction in leaves
Effect of nitrate alone on crop yield
Nitrogen fixation in the presence of nitrate
Establishment of nitrogen fixation
Limitation of nitrogen fixation efficiency by nitrate
Effects on nodule development
Inhibition of nitrogenase activity
How to explain and limit the adverse effects of nitrate?
Complementary role of nitrate on legume nitrogen economy
Under natural conditions
By addition of nitrate as fertilizer
Conclusions
Acknowledgements
References

J.D. Bewley (ed.), Nitrogen and Carbon Metabolism

INTRODUCTION

Bacteria belonging to the genus *Rhizobium* are commonly present in most soils and are able to live in symbiotic association with the roots of legumes. The nodule, the centre of this association, can fix atmospheric nitrogen and this ability places the legumes in an advantageous situation as far as their nitrogen nutrition is concerned. Like other plants, they can also utilize combined nitrogen present in the soils, especially nitrate. It is of interest to compare the respective roles of the nitrogen fixing and nitrate utilization pathways and to determine their relative efficiencies, particularly in relation to the possible improvement of crop productivity at a lower cost.

NITRATE UTILIZATION BY LEGUMES

Nitrate occurs in soils as the result of organic nitrogen oxidation by microorganisms and because it is applied as a nitrogen fertilizer. A profile of nitrate in an Ohio soil, fertilized during the previous crop year shows substantial levels are maintained during a large part of the growing season (1) (Fig.1). Some special features concerning nitrate absorption and its utilization by legumes are summarized below :

Nitrate absorption

Experiments have been carried out using soybeans placed in a mist chamber (2) (3). The absorption kinetics of a dilute nitrate solution (0.5 mM) following an overnight pretreatment with a nitrate-free buffer solution are shown in figure 2. Absorption was found to proceed linearly in both nodulated and nonnodulated soybeans after the first hour, but nodulated soybeans exhibited a nitrate absorption capacity 2.5-fold lower than nonnodulated plants. Wych and Rains (2) have proposed that nitrogen stress is a characteristic of the nodulated plants, which could be responsible

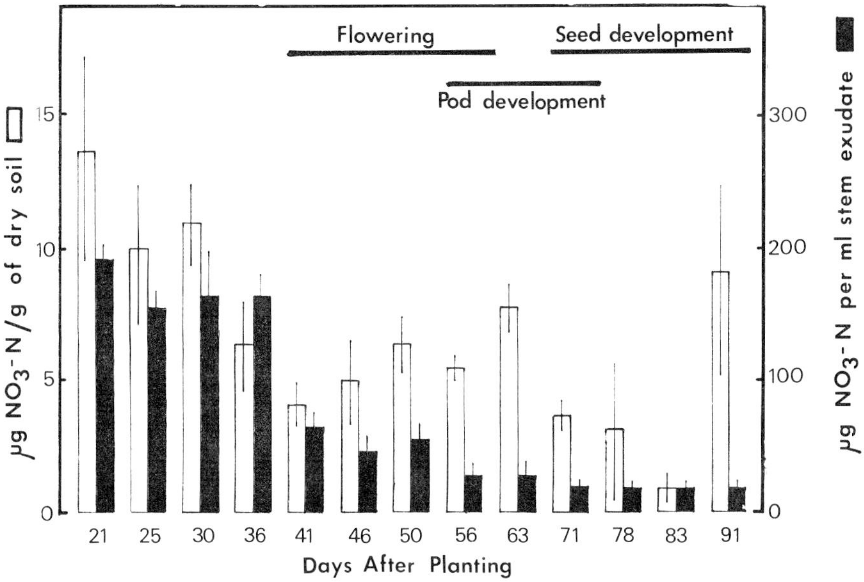

Fig. 1. *Nitrate-N content of soil (depth 20 to 25 cm) and nitrate-N concentration in stem exudates from soybean plants.* Adapted from STREETER J.G. (1).

for their low absorption capacity. A rapid recovery of the nitrate absorption potential of nodulated soybeans was observed when nitrate was resupplied to the culture medium.

That nonnodulated plants have a higher nitrate absorption capacity has also been observed in another type of experiment. Symbiotic nitrogen fixation in the field has been determined by ^{15}N natural abundance determinations (4) (5) (6). Using this technique, it was observed that there is a significantly lower ^{15}N content in nodulating isolines of soybeans (5) and in inoculated soybeans (6) compared with nonnodulating and noninoculated plants, respectively. Since a good correlation has been shown between ^{15}N measurements and C_2H_2 reduction activity, it has been proposed (6) that the difference between ^{15}N content estimates and N yield is due to a lower uptake of soil nitrogen by inoculated soybeans.

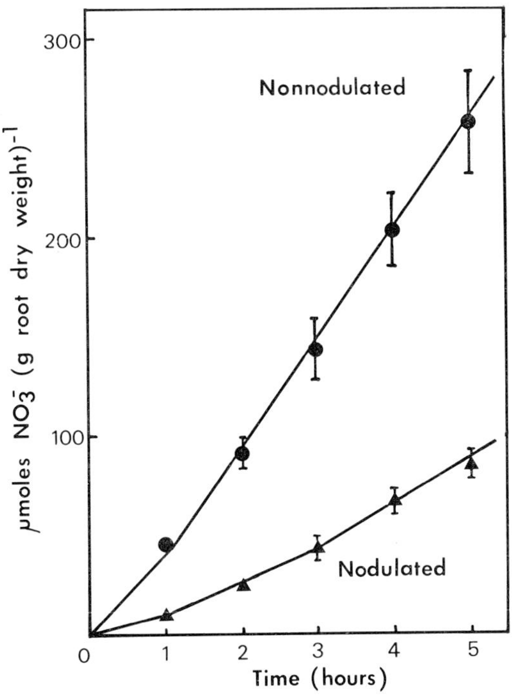

Fig. 2. *Nitrate absorption by nodulated and nonnodulated soybeans, just as flowering begins.* After WYCH R.D. and RAINS D.W. (2).

During the first few days after inoculation with *Rhizobium* , and before nodule emergence, we observed (7) that in hydroponic cultures of *Phaseolus* there is a strong inhibition of root system development. We proposed that this is due to the presence of cytokinins in the medium, which are synthesized when both plants and bacteria are together. Retardation of growth of the inoculated roots continued during the subsequent stages of nodule development. Thus, differences in size of the legume root system could explain the differences in nitrate absorption capacities between nodulated and nonnodulated roots. However, under field conditions, the availability of nitrate in the soil (1) could limit the nitrogen stress period to which nodulating roots are subjected. Thus the morphological changes observed in the laboratory might not be so great in field-grown plants.

The amount of nitrate uptake measured by loss from a liquid medium (8), or by levels in stem exudates (1) (9), varies with the stage of plant development. Nitrate uptake is low in soybeans during the first days after their emergence (8) in hydroponic culture systems. However,field-grown soybeans have substantial levels of nitrate in their sap during their vegetative growth period and these levels correspond with the high nitrate concentrations in the soil (Fig.1). Nitrate uptake decreases during pod and seed development, in spite of its continued availability in the soil (1). Similar observations have been reported with both field-grown and hydroponically-grown soybean plants (9). These observations suggest that supplementing soybean plants with nitrate during the final stages of development will have no effect on nitrogen levels in the pod.

The sites of nitrate utilization

In legumes, both the leaves and the nodulated roots are sites of nitrate utilization.

Nitrate reduction in nodulated roots. Active nitrate reductase has been extracted from field pea roots receiving various doses of nitrate (10). The lateral roots appear to be the most active sites of nitrate reduction therein. There is a rapid induction of nitrate reductase in the presence of nitrate, and the capacity of roots to form this enzyme is well-established in field situations. The reduction of nitrate by field pea roots exposed to different concentrations of this substrate is shown in figure 3. Levels of nitrate reductase in the root are fairly constant over a range of nitrate concentration from 7 to 315 ppm. Shoots, on the other hand, show a dramatic increase in levels of extractable nitrate reductase (11). Root systems in other species of legume (12) have also measurable levels of nitrate reductase and thus have been considered in the scheme of nitrogen metabolism for the considered plants (13).

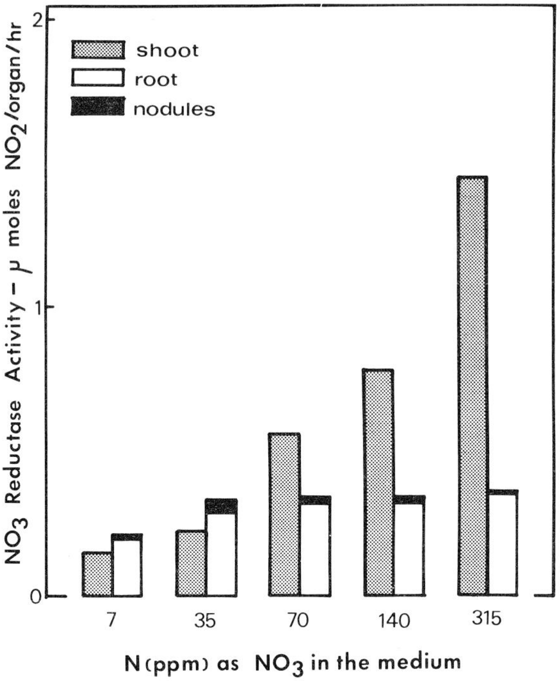

Fig. 3. *Nitrate reductase distribution in shoots, roots and nodules of 28 day-old field pea plants provided with different nitrate levels.* After OGHOGHORIE C.G.O. and PATE J.S.(11)

Nodules themselves can contribute to nitrate reduction (11). An active nitrate reductase has been reported in bacteroids extracted from root nodules of several tropical legumes, including soybean (14). It is interesting to note that in most cases the nodules were induced by rhizobia belonging to the cow pea inoculation group. Recently, the 32 H 1 cow pea strain, which is able to fix nitrogen in culture (15) (16), has been shown to have a constitutive nitrate reductase and it has been used to prepare nitrate reductase deficient mutants (17). Different strains of *Rhizobium japonicum* also exhibit an active nitrate reductase (18) (19) (20), the most common being CC705 (syn. Wisconsin 505) which we used for extraction and purification of the enzyme (20). Nitrate reductase appeared,in that case, to be constitutive and to be present in free-living bacteria cultured on a yeast extract medium, as well as in nodule bacteroids. In both cases, addition of nitrate to the culture media increased the level of nitrate reductase activity. Studies of this nitrate reductase have shown that it is similar in free-living rhizobia and in bacteroids (20). Enzyme activity is present in both a soluble and a particulate form in bacteria and bacteroids (19) (20).

Although only some strains of *Rhizobium* and their bacteroid forms possess a constitutive nitrate reductase, the enzyme can be rapidly induced in other strains by the addition of nitrate. Thus, a large number of rhizobia strains are able to grow on media containing nitrate as sole nitrogen source. However, the inability of seven strains of *Rhizobium lupini* to reduce nitrate in either anaerobic or aerobic conditions (21) has been recently reported.

Two roles could be assigned to this nitrate reduction in nodules :

1. *To contribute to plant nitrogen economy* : This has been proposed by several authors (1) (11). Recently Randall *et al.* (22) (23) detected substantial amounts of reduced ^{15}N in stem exudates of soybean, even though they failed to detect any nitrate reductase in the roots by *in vivo* or *in vitro* assay. They concluded that reduction of $^{15}NO_3$ was occurring in the nodules. Nitrate is able to induce nitrate reductase in the root nodule cytosol of *Pisum sativum* even when it is not possible to detect activity in the bacteroids (12). It has been suggested (23) that the efficiency of nitrate utilization in legumes could be improved by nodule nitrate reductase being prominent during the early stages of vegetative development, when nitrogen fixation is not completely established. Its importance during the last stages of development when leaf nitrate reductase declines (9) (24) (25) seems less likely in view of the reduced ability to absorb nitrate at this time (1) (8) (9).

2. *As an electron acceptor in place of O_2 for bacteroids* : Bacteroids extracted from nodules produced by a strain of *Rhizobium japonicum* (CC705) exhibit a strong nitrate reducing activity. They are able to anaerobically reduce C_2H_2 to C_2H_4 and N_2 to NH_3, with nitrate as the terminal electron acceptor (26). The level of generated ATP is sufficient to provide energy for nitrogenase but the reaction is rapidly limited by nitrite which inhibits nitrogenase (20) (27). In the field, when anaerobic conditions could be induced by soil impermeability or excess of water, nitrate could replace O_2 and could thus permit bacteroid respiration.

Nitrate reduction in leaves. Leaves are considered to be the major site of nitrate reduction in plants (28), and legumes are no exception (8) (10) (24). A greater consideration of this subject is given by Schrader (this volume), but in relation to the topic of this chapter it is of interest to consider some of the conclusions of Harper and Hageman (24) derived from experiments using inoculated soybeans receiving nitrate (8). At the initial flowering and full bloom stages of development of soybean there is marked activity of nitrate reductase which declines from the top to the bottom of the canopy. This activity decreases during the later stages of soybean development and is low at the green bean stage. Similar results were reported with field-grown soybeans by Thibodeau and Jaworski (9) and the rapid drop in nitrate reductase activity prior to pod-fill stage was associated with a low nitrate content in the leaves. In each case a stimulation of enzyme activity has been induced by the addition of exogenous nitrate. These observations suggest an incomplete utilization of nitrate reduction potential under standard field conditions.

Effect of nitrate on crop yield

Assessing the effect of nitrate alone on yield in

legume crops is complicated both by the response of the host plant and of the *Rhizobium* species. Some difficulties are encountered when one attempts to relate the yield of legume crops to their supply of nitrate. For example, under field conditions there is the presence of specific *Rhizobium* strains, which induce nodule formation. But their inoculation potential is variable, and this results in variations in the capacity of induced nodules to fix nitrogen. Hence nitrate utilization can not be determined against a constant value of nitrogen fixation.

Genotype	Nitrogen fertilizer	Seed yield (kg/ha)	Seed protein %	Seed protein (kg/ha)	Nodule f.weight (g/plant)
non-nodulating isoline	Control	2157	31.2	672	----
	NH_4NO_3	2904	39	1149	----
	Urea	3019	39	1176	----
nodulating isoline	Control	2668	37.8	1016	1.3
	NH_4NO_3	3014	39.3	1199	0.6
	Urea	2961	39.7	1178	0.7

Table I. *Effect of nitrogen fertilizers (224 kg N/ha) on soybean yield.* Based on HAM G.E. *et al.* (30).

Utilizing nodulating and nonnodulating soybean isolines, Weber (29) proposed an interesting way to test the efficiency of combined nitrogen alone. The response of nonnodulating soybeans to nitrogen fertilizer was generally considered as linear (29) when there were no adverse conditions during the growing season. Ham *et al.* (30) found a substantial increase in seed yield and protein content in nonnodulating isolines when fertilizer was added compared with the unfertilized control (Table I). A somewhat smaller increase in yield due to fertilizer addition also occurred in the nodulating isoline. Harper (25) also measured responses of soybeans grown in soil plots and under hydroponic conditions. There were slight differences between nodulating and nonnodulating isolines grown in soil supplements with NH_4NO_3. Under hydroponic conditions, on the other hand, nodulating lines responded much better to KNO_3 additions (Table II). This may reflect a more active nitrate reductase activity (25) and a possible role of nitrate as an electron acceptor under essentially anaerobic conditions.

Treatment	Seed yield g/plant	Seed protein %	Seed N g/plant
Soil control			
nodulating isoline	14.8	42.5	1.01
nonnodulating isoline	13.4	38.6	0.83
200 kg N-NH_4NO_3 per ha			
nodulating isoline	16.6	43.5	1.13
nonnodulating isoline	16.2	41.9	1.09
Hydroponics (1.5 mM KNO_3)			
nodulating isoline	10.9	38.7	0.68
nonnodulating isoline	7.8	33.4	0.42

Table II. *Effect of ammonium nitrate on seed yield and protein content of soybean isolines*. After HARPER J.E. (25).

While nitrate is unable to regularly increase the seed yield and protein content of nonnodulated soybeans, dry matter production is always strongly enhanced (29)(30)(31). The simultaneous potassium accumulation reported by Hanway and Weber (32) could be indirectly related to this stimulation. In laboratory experiments, nitrate has been observed to increase the dry weight of non inoculated soybeans in comparison with plants receiving only symbiotic nitrogen (Table III). Nitrogen content was substantially higher in the presence of nitrate and a redistribution of nitrogen from the vegetative parts to the pods and seeds was clearly shown. This confirms previous observations (32) (33). These results are of interest in relation to fodder plants like *Medicago* and *Trifolium*, where nitrogen content and dry matter are simultaneously required for optimum crop yield. In the field situation, it is possible that nitrate application could lead to improved dry matter production in nitrogen-poor soils, in spite of its adverse effect on symbiotic nitrogen fixation.

Age (days)		Dry weight (g/plant)		Tissue N percentage (d.w. basis)	
		Shoots	Seeds + pods	Shoots	Seeds + pods
33	Nod.	6.3±0.1	---	2.8±0.1	---
	Nonnod.	9.1±0.3	---	4.1±0.1	---
53	Nod.	20.0±1.9	2.9±1.1	3.0±0.3	3.8±0.1
	Nonnod.	29.9±3.3	1.6±0.7	3.0±0.1	3.8±0.1
73	Nod.	11.6±0.7	12.7±0.6	2.0±0.1	4.2±0.1
	Nonnod.	30.6±2.1	31.8±2.5	2.7±0.1	4.3±0.1
94	Nod.	6.1±0.6	17.7±2.3	1.1±0.0	4.5±0.1
	Nonnod.	21.8±2.1	58.5±9.1	1.3±0.1	5.1±0.3

Table III. *Dry weights and tissue N percentage of soybeans receiving 6 mM* KNO_3 *(Nonnod.) or N-free solution (Nod.).* Adapted from Wych and Rains (3).

NITROGEN FIXATION IN THE PRESENCE OF NITRATE

Establishment of nitrogen fixation

Nitrogenase activity of nodulated roots is commonly measured by the C_2H_2 reduction technique, which provides rapid and reproducible information on nitrogen fixation (34). However, H_2 evolution from nodules (35) consumes a significant proportion of a plant's energy under field conditions (36), but it is completely inhibited by C_2H_2. Hence, to avoid an overestimation of nitrogen fixation, the theoretical conversion of reduced C_2H_2 into fixed nitrogen will be applied and analyzed critically.

Seasonal patterns of nitrogen fixation by field-grown soybeans have been reported by several authors (9)(24)(30)(34)(37). In soybeans, activity starts at the initial flowering stage which occurs, on average, 4 weeks after sowing. Nitrogen fixation increases rapidly during the flowering period, reaching a maximum during the early and mid-part of pod filling, and then declining sharply with senescence. Weber *et al.* (37) found that about 80 % of the nitrogen was fixed between flowering and the subsequent green bean stage. Similar patterns of nitrogenase activity have been observed in soybeans grown hydroponically (8)(24) and in field grown *Phaseolus* plants (38).

Estimates of fixed nitrogen obtained by different techniques give values for field grown soybeans in the range of 80 to 100 kg N per ha and per season (39) (40). Thus nitrogen fixation furnishes only 30 to 50 % of the plant's requirement for nitrogen, and the rest must be supplied by nitrate available in the soil or as nitrogen-fertilizer. However, contradictory effects on the nitrogen economy of legumes have resulted from nitrate additions.

Limitation of nitrogen fixation efficiency by nitrate

Effects on nodule development. There is an abundance of literature on this subject - see Dart (41)

Gibson (42) and Munns (43) - and it is evident that the different stages of nodule development can be affected by nitrate. The curling of root hairs is considered to be the first step of nodule infection and this is reduced by nitrate application (44) resulting in a significant diminution of *Rhizobium* infection sites. More recently, nitrate has been found to limit both lectin production and bacteria adsorption onto the root hairs (45). The infection threads are also inhibited (43), which also helps to explain why only a few nodules are generally induced in the presence of combined nitrogen. However, in contrast, nitrate has also been reported to stimulate the initial infection processes when added in very low concentrations (41). In addition to the delay in appearance of nodules in the presence of nitrate, a low nodule growing rate has also been observed. This explains the small nodule size and the corresponding low nodule weight which generally is seen in the presence of nitrate (9)(11)(38)(46)(47). In field conditions, this inhibitory effect of nitrate increases with increasing nitrogen-fertilizer supply. When soybeans are cultivated in greenhouses and given a nitrogen-free nutrient medium, the first efficient nodules appear two weeks after sowing (48), a short period in comparison with the four weeks required in the field. This difference cannot be explained on the basis of the optimal greenhouse environment alone.

In contrast, application of low nitrate levels results in a stimulation of nodule size and weight in 28 day-old field peas grown in sand culture (11). Since nitrogen fixation was already well-established this nitrate effect appeared to be synergistic.

Root nodule senescence can be induced by nitrate. Rapid alterations of the nodule fine structure of *Medicago* and *Trifolium* have been observed (49) after a 3- or 8-day treatment with 10.7 mM NH_4NO_3. In soybeans, nodule modifications due to nitrate could be reversible, since plants receiving 7 mM KNO_3 for 6 days recover their nitrogen fixation potential when replaced in a nitrogen-free solution (50). Leghemoglobin, a major nodule constituent (48) can be affected by nitrate (49), since it disappears from pea nodules when the root system is supplied with high nitrate concentrations (100 mM) (12). *Medicago* bacteroid size and nucleic acid content are also affected by 100 mM KNO_3, resulting in a response similar to that induced by stress (51). Under field conditions, then, nitrate concentration can affect nodule senescence processes and could limit nitrogen efficiency.

Inhibition of nitrogenase activity. Although the depressive effect of combined nitrogen on the general nitrogen fixation process is well known, Oghoghorie and Pate (11) clearly demonstrated an inhibition of C_2H_2 reduction by nodules of *Pisum* continuously maintained in the presence of different nitrate concentrations. In hydroponics we observed that a short period of treatment (24 h) with $Ca(NO_3)_2$, 3.5 mM, was sufficient to induce a 60 % inhibition of C_2H_2 reduction by *Phaseolus* nodules (52). There also appears to be a difference in sensitivity to nitrate in relation to nodule age : nitrogenase activity in younger nodules is always less affected by a 24 h nitrate application than that in mature nodules. Depressive effects of nitrate on C_2H_2 reduction were also confirmed in short term experiments using test tubes with liquid culture medium (46) or moistened sand and gravel (53). In greenhouse conditions, however, legume cultures in vermiculite show a lower and variable inhibition rate of C_2H_2 reduction in response to nitrate additions. We consider moisture level to be the major factor affecting the response to combined nitrogen. In the soil situation the normally observed variation in moisture content (1) could be responsible for a discontinuous effect of nitrate upon nitrogen fixation. In addition, the inhibition could be variable within a single plant since any field grown plant contains successive generations of nodules.

solution (50). Leghemoglobin, a major nodule constituent (48) can be affected by nitrate (49), since it disappears from pea nodules when the root system is supplied with high nitrate concentrations (100 mM) (12). *Medicago* bacteroid size and nucleic acid content are also affected by 100 mM KNO_3, resulting in a response similar to that induced by stress (51). Under field conditions, then, nitrate concentration can affect nodule senescence processes and could limit nitrogen efficiency.

Inhibition of nitrogenase activity. Although the depressive effect of combined nitrogen on the general nitrogen fixation process is well known, Oghoghorie and Pate (11) clearly demonstrated an inhibition of C_2H_2 reduction by nodules of *Pisum* continuously maintained in the presence of different nitrate concentrations. In hydroponics we observed that a short period of treatment (24 h) with $Ca(NO_3)_2$, 3.5 mM, was sufficient to induce a 60 % inhibition of C_2H_2 reduction by *Phaseolus* nodules (52). There also appears to be a difference in sensitivity to nitrate in relation to nodule age : nitrogenase activity in younger nodules is always less affected by a 24 h nitrate application than that in mature nodules. Depressive effects of nitrate on C_2H_2 reduction were also confirmed in short term experiments using test tubes with liquid culture medium (46) or moistened sand and gravel (53). In greenhouse conditions, however, legume cultures in vermiculite show a lower and variable inhibition rate of C_2H_2 reduction in response to nitrate additions. We consider moisture level to be the major factor affecting the response to combined nitrogen. In the soil situation the normally observed variation in moisture content (1) could be responsible for a discontinuous effect of nitrate upon nitrogen fixation. In addition, the inhibition could be variable within a single plant since any field grown plant contains successive generations of nodules.

How to explain and limit adverse effect of nitrate ?

Reviews by Dart (41) and Munns (43) indicate the complexity of the nitrate effect on nitrogen fixation and of the difficulty in finding a common explanation for the various phenomena. Nitrate could be influencing nitrogen fixation directly by affecting the root nodules *per se*, or more indirectly by affecting the carbohydrate supply to the centres of nitrogen fixation. Nitrate, or the products of its reduction (nitrite or ammonia), could regulate the function or production of the nodules. The addition of nitrate or ammonia has been associated with lower levels of indoleacetic acid (54), a plant growth substance, thought to be produced by the invading *Rhizobia* and thought to have an important role in nodule formation (41). However, the evidence in favour of this is, at best, indirect. Nitrite, on the other hand, may be directly involved in the functioning of the nodules. As long ago as 1950 Virtanen (55) suggested a nitrite-leghemoglobin complex which could alter the O_2-carrier function of this hemoprotein.

More recently, nitrite has been implicated in the inhibition of nitrogenase activity. Effects have been observed with isolated bacteroids (26), and with crude (20) and purified enzyme preparations (27). Nitrite binds to the MoFe nitrogenase component causing an inhibition of nitrogenase activity. This inhibition appears to be completely reversible. This property parallels the inhibition of nitrogen fixation by nitrate *in vivo* and the rapid restoration of fixation capacities after nitrate has been removed from the medium (50). In this laboratory, a rapid oxidation of ferrous leghemoglobin has also been observed (56) at low nitrite concentrations. For example, figure 4 shows the decline in 541 and 574 nm absorption peaks during deoxygenation and the simultaneous appearance of a 627 nm peak related to the ferric leghemoglobin which is generated. The synthesis of leghemoglobin itself can also be inhibited by combined nitrogen (NH_4NO_3) additions or ammonia alone (57).

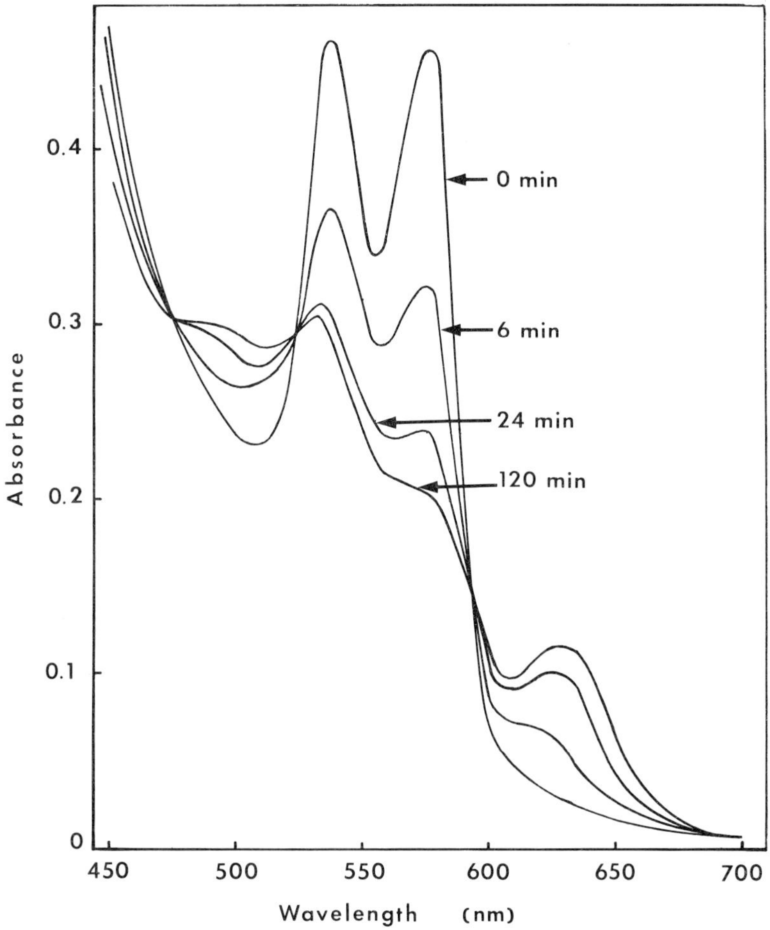

Fig. 4. *Spectral changes of oxy-leghemoglobin (24 µM) in the presence of nitrite (62.5 µM).* After RIGAUD J. and PUPPO A.(56).

The occurrence of nitrite in nodules when plants receive nitrate could be a consequence of nitrate reduction in the bacteroids (see § 1.2.1) and of the induction of the enzyme in the cytosol (12). However, the possible role of nitrite as an inhibitor of nitrogen fixation has recently been questioned by Gibson and Pagan (46). They observed the same levels of inhibition by nitrate in the nodules induced by wild type rhizobia strains as in those induced by nitrate reductase deficient rhizobia. This suggests that other factors are involved in the depression of nitrogenase activity in the nodules.

An alteration in carbohydrate distribution is frequently thought to occur when nodulated plants receive nitrate. The nodule carbon requirement for growth, respiration and export has been estimated to be about 4.1 mg C for each mg of nitrogen fixed (58). Lawrie and Wheeler (59) have pointed out that the role of recently-manufactured photosynthate is for efficient nitrogen fixation in the nodule. They reported a higher rate of accumulation of photosynthetic assimilates in younger infected cells than in those filled with bacteroids. It has been proposed (11) that nitrate, in stimulating shoot nitrate reductase, enhances carbohydrate uptake in the shoot system and causes a carbon shortage in the nodules, resulting in a decline in available energy for nodule growth and nitrogen fixation. In adding tungstate to a soybean culture medium, to replace molybdenum in the Mo-enzymes (60), nitrate reductase was depressed and the nitrate effect on C_2H_2 reduction and nodule growth appeared less pronounced. These results were interpreted as due to an indirect effect of nitrate through its subsequent metabolism (60).

Estimations have been made of the amount of ^{14}C-labeled photosynthate translocated from shoots to the root nodules (47)(61). In the presence of nitrate, nodules receive a lower proportion of photosynthate in comparison to non-nitrate treated plants (61). The depressive effect of nitrate on the photosynthate supply depends on the stage of development of the plant and was found to be particularly marked at full-bloom stage (47). However, it appears to be a general effect of combined nitrogen, since ammonia exerts a similar response, with inhibition rates not statistically different from those reported for nitrate (47). On the other hand, respiratory activity of the nodules involving the utilization of available substrate must also be considered. For example, results reported by Mahon (62) indicate a low sensitivity of root and nodule respiration to the presence of increasing nitrate concentrations which strongly depressed C_2H_2 reduction.

Several suggestions have been made on how to alleviate a possible photosynthate deprivation. In laboratory experiments, addition of organic substrates like glucose (63) or sucrose (64) can partially limit nitrate effects on nodulation and C_2H_2 reduction. But in experiments using NH_4Cl, a 30 % recovery of C_2H_2 reduction was also observed (64) making the results difficult to interpret in terms of nitrate reductase competition.

Since photosynthate is considered to be the general limiting factor for nitrogen fixation, CO_2 enrichment techniques have been used to increase the rate of photosynthesis (65). Increasing CO_2 concentrations supplied to *Pisum* shoots were unable to overcome the depressive effect of 20 mM nitrate on C_2H_2 reduction, although more assimilates were available for the root nodules (12). In other experiments, foliar application of nitrate stimulated leaf nitrate reductase without affecting the nitrogenase activity of nodules, pointing out a sufficiency of photosynthate for the nodules (12).

It can be concluded that nitrate is able to induce a competition for carbohydrate utilization, which could affect nodule development as well as nitrogenase activity. The latter could be lowered by a more direct effect of nitrate, through its reduction products.

Complementary role of nitrate on legume nitrogen economy

Under natural conditions. Nitrogenase activity has been detected 3 or 4 weeks after seed emergence, at a time when the nitrogen content of the seed is exhausted and when nitrogen nutrition is satisfied by combined nitrogen absorbed from the soil. Thus nitrate can play a major role in limiting the nitrogen stress during early seedling development. The capacity for its absorption is maximal during the first stages of development and a very high level of nitrate reduction occurs in the leaves. Figure 5 summarizes a general scheme illustrating the successive utilization of nitrate and atmospheric nitrogen in legume nutrition. The rapid decline in leaf nitrate reductase

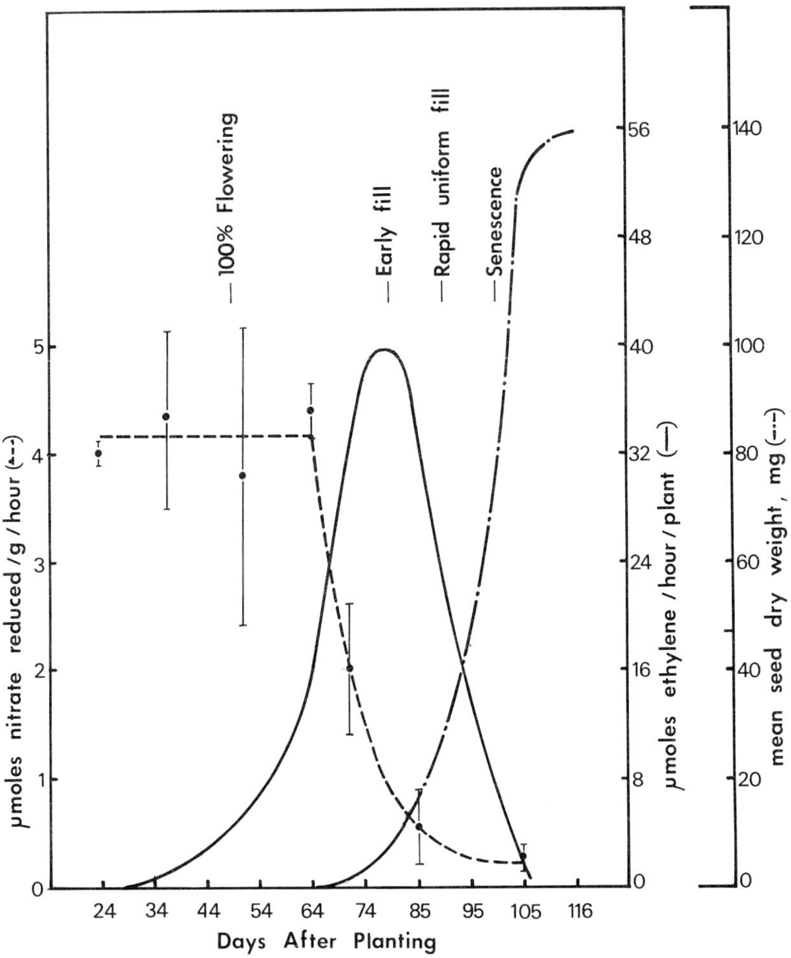

Fig. 5. *Relationships between nitrate reductase activity, nitrogen fixation (C_2H_2 reduction), and seed development in field-grown soybeans.* After THIBODEAU P.S. and JAWORSKI E.G. (9).

activity after the early pod fill stage is associated with a rapid drop in nitrate absorption capacity, and is accompanied by an increasing rate of nitrogen fixation. Nitrogenase activity is maximal during the pod filling period and is followed by a rapid drop in activity which is related to nodule senescence. Seed maturity is not completed by the time nitrogen fixation ceases, and a major part of the subsequent demand for nitrogen is satisfied by nitrogen redistribution from vegetative parts (66) (67). It is difficult to assign a role to nitrate during this later period under field conditions for the level of available soil nitrate is low (Fig.1), soil moisture is inadequate and nitrate reduction is not appreciable. It is necessary to take this particular situation into account when considering aspects of nitrogen fertilizer supply at later stages of development.

By addition of nitrate as fertilizer. To improve legume nitrogen nutrition, by complementing symbiotic nitrogen fixation, two periods within the growing season must be considered : the early stages of development and the reproductive period.

During the first stages of development, provided that the absorption capacity is sufficient and that there are substantial levels of nitrate reductase, nitrate can play a major role in the nitrogen nutrition of legumes. To avoid its depressive effect on nodulation, it appears that the best time of application is after nodule initiation (68). Low levels of combined nitrogen stimulate an increase in nodule number and growth in *Vigna* (68) or *Pisum* (11) and nitrogen starvation is reduced. Dry matter production and C_2H_2 reduction are enhanced in *Medicago* under these conditions (69). Nevertheless the extent of stimulation is affected by various factors, such as environmental conditions, plant genotypes and bacterial strains (68)(69)(70). Similar variations in the degree to which nitrogen fixation is enhanced (ranging 0 to 100 %) have been found to occur in *Trifolium* in test tube experiments (50).

Treatment mM NO_3^-	Yield g/plant	Protein %	N g/plant
0.0 → 0.0	7.2	41.4	0.48
0.0 → 0.75	9.6	40.8	0.63
0.0 → 7.5	11.9	42.7	0.81
0.75 → 0.75	11.2	40.7	0.73
0.75 → 7.5	15	45.1	1.09
7.5 → 7.5	12.9	45.1	0.93
7.5 → 0.75	14.2	40.5	0.92

The nutrient treatment switches were made after full bloom stage.

Table IV. *Effect of nitrate treatments on yield and N content of hydroponically grown soybeans.* After HARPER J.E. (25).

Table IV shows the effects of changing the concentration of nitrate in the medium in which soybeans were grown at the time of full bloom. Soybeans which first received 0.75 mM nitrate and were transferred to 7.5 mM nitrate exhibited a higher seed yield than those plants continuously maintained on 7.5 mM nitrate. These differences are mainly due to variations of nitrogen fixation efficiency in direct relation with the level of combined nitrogen added. Thus, low levels of nitrate supplied early allowed for a better final yield by stimulating root development (71) and general growth (72)(73)(74) without strongly affecting the symbiotic system. Hence the basal combined nitrogen level (30 ppm) used during growth of the cow pea in greenhouse conditions (73) could be used in the field situation where soil nitrogen is low. It is important to note, however, that the response to nitrate is also affected by soil moisture levels (25).

Another critical period for legume nitrogen nutrition is from pod-fill to maturity, when nitrogen fixation is declining (Fig.5) and nitrogen demand is high. The length of the filling stage in legumes is, in general, quite long (and particularly in soybeans) and hence the critical period as far as final yield is concerned is considerable. Addition of nitrogen fertilizer to nodulated cow pea from mid-pod-fill to maturity delays senescence of the nodules and gives more pods and peduncles per plant (Table V).

Treatment (ppm N)	Peduncles/ plant	Pods/ plant	Seeds/ pod	Mean seed wt (mg)
Control (30)	28.7	72	10.6	85.0
60	31.8	80.2	9.0	86.2
120	36.8	89	9.6	86.3
240	33.7	83.9	8.9	88.2

Table V. *Effect of combined nitrogen applied from mid-pod-fill to maturity on the components of seed yield from nodulated cow pea. (The control received 30 ppm N throughout the season).* Adapted from DART *et al.* (73).

Although seed yield is not significantly increased the plant yield is. The extent of redistribution of material from the vegetative parts to the seeds during the seed maturation period (32)(33) is related to the senescence processes (33)(66)(75). If an external nitrogen supply results in more material being available for redistribution within the plant, this might be expected to delay leaf senescence, thus prolonging efficient photosynthesis (40)(65).

The net result should be high seed or pod yield. However, both cow peas (73) and soybeans (33) do not respond to fertilization during this reproductive period. This indicates that other factors are involved in the senescence and maturity processes (76). The low absorption potential of roots during the pod-fill stage (1)(9) which is associated with a poor level of leaf nitrate reductase (9)(24) and a low leaf carbohydrate level, could be responsible for the inadequate utilization of nitrate added to the soil.

CONCLUSIONS

Plants dependent on atmospheric nitrogen as the sole source of nitrogen show nitrogen deficient systems early in the growth period before the nodules are fully developed, and the leaves tend to senesce before pod filling is completed. When nitrate is supplied early, in moderate amounts, it enhances the development of nodules, but in excessive amounts it inhibits nodule development. At later stages of development, both the absorption of nitrate and its reduction are decreased by natural senescent processes occurring in the plant and result in a poor plant response to added nitrogen fertilizer. At these late stages foliar application of nitrogen (nitrate, ammonia or urea) fertilizer will sometimes delay the onset of senescence. Research suggests that varieties of plants which possess high levels of nitrate reductase late into the pod maturity stage should be selected for in breeding programs. Preliminary experiments suggest that such plants have higher potentials for nitrate utilization at the pod-fill stage.

AKNOWLEDGMENTS

The author would like to thank Professors J.D. Bewley and Ann Oaks for their suggestions in the preparation of the manuscript.

REFERENCES

1. STREETER J.G., Nitrogen nutrition of field-grown soybean plants.I.- Seasonal variations in soil nitrogen and nitrogen composition of stem exudate. Agron. J. 64: 311-314, 1972
2. WYCH R.D. and D.W. RAINS, Simultaneous measurement of nitrogen fixation estimated by the acetylene-ethylene assay and nitrate absorption by soybeans. Plant Physiol. 62: 442-448, 1978
3. WYCH R.D. and D.W. RAINS, NO_3^- absorption and C_2H_2 reduction by soybeans during reproductive development. Physiol. Plant. 47: 200-204, 1979
4. BARDIN R., A.M. DOMENACH and A. CHALAMET, Rapports isotopiques naturels de l'azote. II. Application à la mesure de la fixation symbiotique de l'azote *in situ*. Rev. Ecol. Biol. Sol 14: 395-402, 1977
5. KHOL D.H., G. SHERER and J.E. HARPER, The natural abundance of ^{15}N nodulating and non-nodulating isolines of soybeans stable isotopes : Proceedings of the IIIrd Int. Cont. (1978). Acad. Press New York, in press
6. AMARGER N., A. MARIOTTI, F. MARIOTTI, J.C. DURR, C. BOURGUIGNON and B. LAGACHERIE, Estimate of symbiotically fixed nitrogen in field grown soybeans using variations in ^{15}N natural abundance. Plant and Soil 52: 269-280, 1979
7. PUPPO A. and J. RIGAUD, Cytokinins and morphological aspects of French-bean roots in the presence of *Rhizobium*. Physiol. Plant. 42: 202-206, 1978
8. HARPER J.E., Seasonal nutrient uptake and accumulation patterns in soybeans. Crop Sci. 11: 347-350, 1971
9. THIBODEAU P.S. and E.G. JAWORSKI, Patterns of nitrogen utilization in the soybean. Planta 127: 133-147,1975
10. WALLACE W. and J.S. PATE, Nitrate reductase in the field pea (*Pisum arvense* L.). Ann. Bot. 29: 655-671, 1965

11. OGHOGHORIE C.G.O. and J.S. PATE, The nitrate stress syndrome of the nodulated field pea (*Pisum arvense* L.). Techniques for measurement and evaluation in physiological terms. Plant and Soil, Spec. vol.185-202. Lie T.A. and E.G. Mülder eds. Nijhoff. The Hague 590 p., 1971

12. CHEN P.C. and D.A. PHILLIPS, Induction of root nodule senescence by combined nitrogen in *Pisum sativum* L. Plant Physiol. 59: 440-442, 1977

13. OGHOGHORIE C.G.O. and J.S. PATE, Exploration of the nitrogen transport system of a nodulated legume using ^{15}N. Planta 104: 35-39, 1972

14. CHENIAE G. and H.J. EVANS, Physiological studies on nodule nitrate reductase. Plant Physiol. 35: 454-462, 1960

15. PAGAN J.D., J.J. CHILD, W.R. SCOWCROFT and A.H. GIBSON, Nitrogen fixation by *Rhizobium* cultured on a defined medium. Nature 256: 406-407, 1975

16. KURZ W.G.W. and T.A. LARUE, Nitrogenase activity in rhizobia in absence of plant host. Nature 256: 407-408, 1975

17. PAGAN J.D., W.R. SCOWCROFT, W.F. DUDMAN and A.H. GIBSON, Nitrogen fixation in nitrate reductase-deficient mutants of cultured rhizobia. J. Bacteriol. 129: 718-723, 1977

18. BERGERSEN F.J., Nitrate reductase in soybean root nodules. Biochim. Biophys. Acta 52: 206-207, 1961

19. LOWE R.H. and H.J. EVANS, Preparation and some properties of a soluble nitrate reductase from *Rhizobium japonicum*. Biochim. Biophys. Acta 85: 377-389, 1964

20. KENNEDY I.R., J. RIGAUD and J.C. TRINCHANT, Nitrate reductase from bacteroids of *Rhizobium japonicum* : Enzyme characteristics and possible interaction with nitrogen fixation. Biochim. Biophys. Acta 397: 24-35, 1975

21. MANHART J.R. and P.P. WONG, Nitrate reductase activities of rhizobia and the correlation between nitrate reduction and nitrogen fixation. Can. J. Microbiol. 25: 1169-1174, 1979

22. RUSSEL W.J., D.R. JOHNSON and D.D. RANDALL, Nitrate reduction in soybean nodules and leaves in relation to N_2-(C_2H_2) fixation. Agron. Abstr. 1974 Annual Meeting p. 76

23. RANDALL D.D., W.J. RUSSEL and D.R. JOHNSON, Nodule nitrate reductase as a source of reduced nitrogen in soybean *Glycine max*. Physiol. Plant. 44: 325-328,1978

24. HARPER J.E. and R.H. HAGEMAN, Canopy and seasonal profiles of nitrate reductase in soybeans (*Glycine max* L. Merr.). Plant Physiol. 49: 146-154, 1972

25. HARPER J.E., Soil and symbiotic requirements for optimum soybean production. Crop Sci. 14: 255-260, 1974

26. RIGAUD J., F.J. BERGERSEN, G.L. TURNER and R.M. DANIEL, Nitrate dependent anaerobic acetylene reduction and nitrogen fixation by soybean bacteroids. J. Gen. Microbiol. 77: 137-144, 1973

27. TRINCHANT J.C. and J. RIGAUD, Nitrite inhibition of nitrogenase from soybean bacteroids. Arch. Microbiol. 124: 49-54, 1980

28. BEEVERS L. and R.H. HAGEMAN, Nitrate reduction in higher plants. Ann. Rev. Plant Physiol. 20: 495-522, 1969

29. WEBER C.R., Nodulating and non nodulating soybean isolines. II. Response to apply nitrogen and modified soil conditions. Agron. J. 58: 46-49, 1966

30. HAM G.E., R.J. LAWN and W.A. BRUN, Influence of inoculation, nitrogen fertilizers and photosynthetic source-sink manipulations on field-grown soybeans.In: Symbiotic nitrogen fixation in plants 239-253.Nutman P.S. ed., Cambridge Univ. Press 594 p., 1976

31. HANWAY J.J. and C.R. WEBER, Dry matter accumulation in soybean (*Glycine max* L. Merril) plants as influenced by N.K.P. fertilization. Agron.J. 63: 365-368, 1971

32. HANWAY J.J. and C.R. WEBER, Accumulation of N,P and K by soybean (*Glycine max* L. Merril) plants. Agron. J. 63: 406-408, 1971

33. EGLI D.B., J.E. LEGETT and W.C. DUNCAN, Influence of nitrogen-stress on leaf senescence and N redistribution in soybeans. Agron. J. 70: 43-47, 1978

34. HARDY R.W.F., R.D. HOLSTEN, E.K. JACKSON and R.C. BURNS, The acetylene-ethylene assay for N_2 fixation : laboratory and field evaluation. Plant Physiol. 43: 1185-1207, 1968

35. HOCH G.E., H.N. LITTLE and R.H. BURRIS, Hydrogen evolution from soybean root nodules. Nature 119: 430-431, 1952

36. SCHUBERT K.R. and H.J. EVANS, Hydrogen evolution a major factor affecting the efficiency of nitrogen fixation in nodulated symbionts. Proc. Nat. Acad.Sci. USA 73: 1207-1211, 1976

37. WEBER D.E.,B.E. CALDWELL, C.SLOGER and H.G. VEST, Some USDA studies on the soybean rhizobium symbiosis. In: Plant and Soil. Spec. vol. 293-304 Lie T.A. and E.G. Mulder eds., Nijhoff. The Hague 590 p., 1971

38. FRANCO A.A., J.C. PEREIRA and C.A. NEYRA, Seasonal patterns of nitrate reductase and nitrogenase activities in *Phaseolus vulgaris* L. Plant Physiol. 63: 421-424, 1979

39. STEWART W.D.P., Nitrogen fixation in plants. The Athlone Press University of London. London, 1966

40. HARDY R.W.F. and U.D. HAVELKA, Nitrogen fixation research : a key to world food ? Science 188: 633-643, 1975

41. DART P.J., The infection process. In : The biology of nitrogen fixation 382-429 Quispel A. ed. North-Holland publishing Comp. Amsterdam 769 p., 1974

42. GIBSON A.H., The control of dinitrogen assimilation by nodulated legumes. In: Mechanisms of regulation of plant growth. Bieleski R.L., A.R.Ferguson, M.M. Cresswell, Royal Society of New Zealand, Wellington, 13-22, 1974

43. MUNNS D.N., Mineral nutrition and the legume symbiosis p. 353-391. In : A treatise on dinitrogen fixation. Section IV. Agronomy and ecology. Hardy R.W.F. and A.H. Gibson eds. Wiley, New York, 527 pp., 1977

44. MUNNS D.N., Nodulation of *Medicago sativa* in solution culture. III. Effects of nitrate on root hairs and infection. Plant and Soil 29: 33-47, 1968

45. DAZZO F.B. and W.J. BRILL, Regulation by fixed nitrogen of host symbiont recognition in the *Rhizobium*-clover symbiosis. Plant Physiol. 62: 18-21, 1978

46. GIBSON A.H. and J.D. PAGAN, Nitrate effects on the nodulation of legumes inoculated with nitrate-reductase-deficient mutants of *Rhizobium*. Planta 134: 17-32, 1977

47. LATIMORE M., J.J. GIDDENS and D.A. ASHLEY, Effect of ammonium and nitrate nitrogen upon photosynthate supply and nitrogen fixation by soybeans. Crop Sci. 17: 399-404, 1977

48. BERGERSEN F.J. and D.J. GOODCHILD, Cellular location and concentration of leghaemoglobin in soybean root nodules. Aust. J. Biol. Sci. 26: 741-756, 1973

49. DART P.J. and V.H. MERCER, The influence of ammonium nitrate on the fine structure of nodules of *Medicago tribuloïdes* Desr. and *Trifolium subterraneum* L. Arch. Mikrobiol. 51: 233-257, 1965

50. GIBSON A.H., Recovery and compensation by nodulated legumes to environmental stress. In : Symbiotic nitrogen fixation in plants 239-253. Nutman P.S. ed. Cambridge Univ. Press 584 p. 1976

51. PAAU A.S. and J.R. COWLES, Effect of induced nodule senescence on parameters related to dinitrogen fixation, bacteroid size and nucleic acid content. J. Gen. Microbiol. 111: 101-107, 1979

52. RIGAUD J., Effet des nitrates sur la fixation d'azote par les nodules de Haricot (*Phaseolus vulgaris* L.). Physiol. Vég. 14: 297-308, 1976

53. KAMBERGER W., Regulation of symbiotic nitrogen fixation in root nodules of alfalfa (*Medicago sativa*) infected with *Rhizobium meliloti*. Arch. Microbiol. 115: 103-108, 1977

54. TANNER J.W. and I.C. ANDERSON, External effect of combined nitrogen on nodulation. Plant Physiol. 39: 1039-1043, 1964

55. VIRTANEN A.I., Microbiology and chemistry of symbiotic nitrogen fixation. Proc. VII Int. Bot. Congr. Stockholm pp. 156-159, 1950

56. RIGAUD J. and A. PUPPO, Effect of nitrite upon leghemoglobin and interaction with nitrogen fixation. Biochim. Biophys. Acta 497: 702-706, 1977

57. BISSELING T., R.C. VAN DEN BOS and A. VAN KAMMEN, The effect of ammonium nitrate in the synthesis of nitrogenase and the concentration of leghemoglobin in pea root nodules induced by *Rhizobium leguminosarum*. Biochim. Biophys. Acta 539: 1-11, 1978

58. MINCHIN F.R. and J.S. PATE, The carbon balance of a legume and the functional economy of its root nodules. J. Exp. Bot. 24: 259-271, 1973

59. LAWRIE A.C. and C.T. WHEELER, The supply of photosynthetic assimilates to nodules of *Pisum sativum* L. in relation to the fixation of nitrogen. New Phytol. 72: 1341-1344, 1973

60. HARPER J.E. and J.C. NICHOLAS, Nitrogen metabolism of soybeans. I. Effect of tungstate on nitrate utilization, nodulation and growth. Plant Physiol. 62: 662-664, 1978

61. SMALL J.G.C. and O.A. LEONARD, Translocation of ^{14}C-labeled photosynthate in nodulated legumes as influenced by nitrate nitrogen. Amer. J. Bot. 56: 187-194, 1969

62. MAHON J.D., Respiration and the energy requirement for nitrogen fixation in nodulated pea roots. Plant Physiol. 60: 817-821, 1977

63. WONG P., Effects of nitrate and carbohydrates on nitrogen fixation activity of legume root nodules. Plant Physiol. 59 S: 50, 1977

64. HOUWAARD D., Effect of combined nitrogen on symbiotic nitrogen fixation in pea plants. Thesis,Agricultural University, Wageningen, 98 p. 1979

65. HARDY R.W.F. and U.D. HAVELKA, Photosynthate as a major factor limiting nitrogen fixation by field-grown legumes with emphasis on soybeans. In : Symbiotic nitrogen fixation in plants. Nutman P.S. ed. pp. 421-439, 1976

66. SINCLAIR T.R. and C.T. DEWITT, Analysis of the carbon and nitrogen limitations to soybean yield. Agron. J. 68: 319-324, 1976

67. LAWN R.J. and W.A. BRUN, Symbiotic nitrogen fixation in soybeans. I. Effect of photosynthetic source-sink manipulations. Crop Sci. 14: 11-16, 1974

68. PATE J.S. and P.J. DART, Nodulation studies in legumes. IV. The influence of inoculum strain and time of application of ammonium nitrate on symbiotic response. Plant and Soil 15: 329-346, 1961

69. HOGLUND J.H., Bimodal response by nodulated legumes to combined nitrogen. Plant and Soil 39: 533-545, 1973

70. GIBSON A.H., Consideration of the growing legume as a symbiotic association. Proc. Ind. Nat. Sci. Acad. 40 B: 741-767, 1974

71. HARPER J.C. and R.L. COOPER, Nodulation response of soybeans (*Glycine max* L. Merr.). Application rate and placement of combined nitrogen. Crop Sci. 11: 438-440, 1971

72. MAHON J.D. and J.J. CHILD, Growth response of inoculated peas (*Pisum sativum*) to combined nitrogen. Can. J. Bot. 57: 1687-1693, 1979

73. DART P.J., P.A. HUXLEY, A.R. EAGLESHAM, F.R. MINCHIN, R.J. SUMMERFIELD and J.M. DAY, Nitrogen nutrition of cow pea (*Vigna unguiculata*). II. Effects of inorganic nitrogen on growth and yield of nodulated and non nodulated plants. Expl. Agric. 13: 241-252, 1977

74. SUMMERFIELD R.J., P.A. HUXLEY, P.J. DART and A.P.HUGUES, Some effects of environmental stress on seed yield of cow pea (*Vigna unguiculata* L.) Walp cv. prima. Plant and Soil 44: 527-546, 1976

75. SINCLAIR T.R. and C.T. de WIT, Comparative analysis of photosynthate and nitrogen requirements in the production of seeds by various crops. Science 189: 565-567, 1975

76. LEOPOLD A.C. and P.E. KRIEDEMANN, Plant growth and development. McGraw-Hill Book Co. New York 545 p., 1975

CHAPTER 3

NITRATE UPTAKE, REDUCTION AND TRANSPORT IN THE WHOLE PLANT

L.E. Schrader and R.J. Thomas
Department of Agronomy, University of Wisconsin-Madison, Madison, WI. 53706

Introduction
Nitrate Uptake
Nitrate Assimilation
  Intracellular Localization of Enzymes
  Intercellular Localization of Enzymes
  Electron Donors for Nitrate Assimilation
  Relationship of Nitrate Assimilation and Photosynthesis
  Regulation of Nitrate Reductase
    a) Regulation of amount of enzyme
    b) Regulation of nitrate reductase activity
  Relationship of NR to Grain Protein and Yield
Transport of Reduced Nitrogen
  Composition of Xylem and Phloem Saps in Plants Grown on Nitrate
  Phloem Loading and Transport of Nitrogenous Solutes
  Transport of Nitrogen as a Possible Limiting Factor for Productivity
Summary
References

J.D. Bewley (ed.), Nitrogen and Carbon Metabolism

INTRODUCTION

The major form of inorganic nitrogen available to higher plants in most soils is nitrate because fertilizer N and ammonia from ammonification of organic matter are readily oxidized to nitrate-N through nitrification by *Nitrosomonas* and *Nitrobacter*. It is estimated that only about 50% of the fertilizer N that is applied to crops is normally recovered by the crop during the year of application (1-4).

Utilization of nitrate by higher plants includes several processes, any one of which could limit nitrate assimilation (5-7). The first control point for nitrate assimilation is the uptake system (8). Nitrate absorbed by roots may be reduced by nitrate reductase (NR) in the roots, accumulated in storage sites (e.g., vacuoles), or transported via the xylem to the shoot. Nitrate accumulated or stored in vacuoles or other sites in which NR does not occur may be translocated at a later time to a site of reduction (9,10) and thereby serve as a reserve pool of nitrate; or nitrate may remain in the storage sites until harvest (4,7,11). The most limiting enzyme involved in assimilating nitrate into amino acids is thought to be NR (7,12,13). Some species reduce considerable amounts of nitrate-N in their roots whereas others reduce most of it in the shoots (13). Ammonia formed during reduction of nitrate in roots or shoots must be combined with carbon before it can be transported to other plant parts or cells (14). The formation of amino acids and other forms of reduced N is regulated (15) and thus may be a control point. The loading and transport of these nitrogenous compounds may also be controlled or limited in some species.

Each process in the uptake, distribution, reduction, and incorporation of N into organic forms is under genetic influence (16); hence one might expect to find genetic variability for each of these traits or processes. Identification of this genetic variability and knowledge of the inheritance of these traits should permit breeders to select genotypes possessing an increased capacity to convert soil N to plant protein. A better understanding of the physiology and biochemistry of nitrate assimilation in plants is needed, however, to help identify the process(es) that is most limiting.

Because of space limitations, comprehensive coverage of even the most recent literature in all these areas will be impossible. Several recent review articles will be cited for the reader who desires a more extensive coverage of these topics.

NITRATE UPTAKE

The uptake of nitrate from the environment has been studied in several important crop plants, and many reviews deal with the characteristics of the uptake system involved (5-7,17,18). Here we will briefly mention the salient points of the physiological capacity of plants to absorb nitrate from the environment and indicate areas where progress may be made in improving the efficiency of nitrate uptake by crop plants.

/Some crop plants previously grown without nitrate show a lag or induction phase when given nitrate. The rate of uptake is initially slow, but increases steadily until an "accelerated" rate is attained (6). This pattern of uptake is in contrast to that of other ions which generally exhibit linear rates during the initial stages of uptake. The phenomenon of an apparent "induction" phase may not be of general occurrence as liquid cultures of perennial ryegrass (Lolium perenne) when given nitrate after a period of N starvation did not show a lag phase in nitrate uptake (19).

Uptake appears to require energy (5,20) although an anion ATPase has not been conclusively demonstrated for nitrate uptake in higher plants. Inhibitors of RNA and protein synthesis have been shown to decrease nitrate uptake suggesting a dependence on the continual synthesis of a protein or postulated "permease" (6). The observed enhancement of nitrate uptake by light may be partly the result of an increased supply of energy (20) or an increased supply of assimilates from the shoot. Ben Zioni et al. (21,22) proposed a model in which nitrate uptake is regulated by nitrate reduction and malate production in the shoot. Malate is postulated to be transported along with K^+ to the root where malate is oxidized and decarboxylated to form bicarbonate which is then exchanged for external nitrate. While evidence has been published to support this scheme (e.g., 22-24), there is still no direct evidence that K^+malate is involved in the control of nitrate uptake. Total organic acids, rather than malate, have usually been measured. Deane-Drummond et al. (25) observed a decrease in root malate concentrations after decapitation of barley shoots indicating that the increase in malate (and presumably nitrate uptake) in root tissue during the day was dependent on the downward translocation of malate from the shoot. It would be of interest to know if the observed diurnal rhythm of nitrate uptake (26) is accompanied by similar fluctuations in malate transport. The transfer of the negative charge from malate to the external solution via bicarbonate in exchange for nitrate is thought to be related to changes in internal pH (27) but further work is required to substantiate this model. It is doubtful that the Ben Zioni model (21) is applicable to all crop plants as, for example, some C_3 plants such as tomato appear to retain malate in the shoot rather than transport it to the root (28). In C_4 plants such as maize, difficulties can be expected in differentiating between the effects of nitrate metabolism on leaf malate content from those associated with malate production during C_4 metabolism. The model is further complicated by the observation that malate may be utilized rather than synthesized during nitrate reduction in maize leaves, via malate oxidation which has been shown to provide NADH for nitrate reduction (29).

The decrease in nitrate uptake following decapitation of the shoot and a partial restoration of the uptake rate to that of the intact plant by glucose feeding indicates the close dependence of the uptake process on a continual supply of assimilates from the shoot (6,30). We require more information on the role of assimilates in nitrate uptake and ideally in systems where the effects on uptake and reduction can be separated. Doddema & Otten (31), using a mutant of Arabidopsis which absorbs nitrate but has no NR, have shown that there is a correlation between nitrate uptake and root malate content. Further use of similar mutants in other crop species will provide an insight into the role of the shoot in the regulation of nitrate uptake.

Ambient ammonium and high internal concentrations of nitrate have been shown to decrease nitrate uptake (5,18,32) although the precise mechanisms are not known. Generally there is a parallel "induction" of nitrate uptake and reduction but again little is known about the exact nature of the relationships between the two (5,7,17).

From kinetic data of nitrate uptake there appear to be at least two mechanisms involved. Nitrate uptake rates measured over a range of low nitrate concentrations can be described by a simple rectangular hyperbola. At higher concentrations the rate continues to increase above the apparent saturation rate observed at lower concentrations (5). Work with mutants of Arabidopsis also indicates that there are two or more independent mechanisms which presumably are genetically determined (33). These complex kinetic data for nitrate uptake are similar to those reported for other ions showing bi- or multiphasic kinetics (34,35). It is difficult to assess what these kinetic data mean with respect to the field situation as the concentrations of available nitrate in the soil depend on various environmental conditions, but it is likely that uptake usually occurs over a range of low nitrate concentrations. Clement et al. (36) showed little change in nitrate uptake rates of ryegrass (Lolium) over a wide range of nitrate concentrations likely to be found in the field. These results suggest that plants may have a mechanism to maintain uptake rates at low ambient nitrate concentrations.

One of the important goals for the future improvement of nitrate uptake by crop plants is to determine the extent of the genetic variability in a species' capacity to efficiently utilize the available nitrate in the soil. Selection for variants which have higher uptake rates could result in more rapid recovery of soil N before losses to the environment occur. This could improve the current figure of a 50% efficiency of fertilizer utilization (4). Genotypic differences in nitrate uptake have been reported in maize (37) and wheat (5). An increase in uptake rate per se may not be beneficial in terms of crop productivity if the increase is a result of increased root growth at the expense of shoot growth. Ideally a lowering of the apparent K_m for uptake (increased affinity) is desired as this would be unlikely to alter the root:shoot ratio. There is little evidence for intraspecific differences in the apparent K_m (or half-saturation constant) for nitrate uptake amongst higher plants although inter-specific differences have been noted (e.g., 5,38). In algae intraspecific differences have been observed. Clones of the same species of marine phytoplankton had different K_m values for nitrate uptake (39). Those isolated from nutrient-poor waters had higher affinities (lower K_m's) than clones of the same species isolated from nutrient-rich waters.

Research to improve nitrate uptake has been hampered by the lack of a suitable isotope of N. Currently the use of $^{36}ClO_3$ as an analogue of nitrate in uptake studies is being tested as a tool for the screening of plants for increased uptake efficiencies (Rhodes, Zabala & Filner, personal communication). A disadvantage of this technique is that it is likely to be destructive due to chlorate toxicity.

A problem which is likely to be encountered with any improved capacity for nitrate uptake is that of nitrate accumulation (7). Friedrich et al. (9) showed that the accumulated nitrate in roots and stems of maize is available for assimilation especially if the supply of ambient nitrate decreases during grain filling. This taken together with the finding that most of the N in the ears of maize is absorbed by the plant prior to silking (10) indicates that, in these plants, the most N-efficient will be those which can absorb most of their N early during vegetative growth and efficiently remobilize any accumulated nitrate. In plants which absorb nitrate during reproductive growth (40), it will be important to determine if this N is remobilized efficiently into the harvested plant parts.

Physiologists need to identify the limiting and controlling factors involved in nitrate accumulation and remobilization so that any genetic variations in these processes, such as those noted for nitrate accumulation

(37,41) can be exploited via breeding programs.

NITRATE ASSIMILATION

Nitrate assimilation includes both reductive and non-reductive processes in converting nitrate-N to amino-N. The reactions of this pathway have been discussed in several recent reviews (40,42-46). The four enzymes involved are NR, nitrite reductase (NiR), glutamine synthetase (GS), and glutamate synthase (GOGAT). Three of these reactions are reductive, and a total of 10 electrons must be provided, as shown below, for the assimilation of a nitrate ion to glutamate. The provision of these electrons will be discussed later.

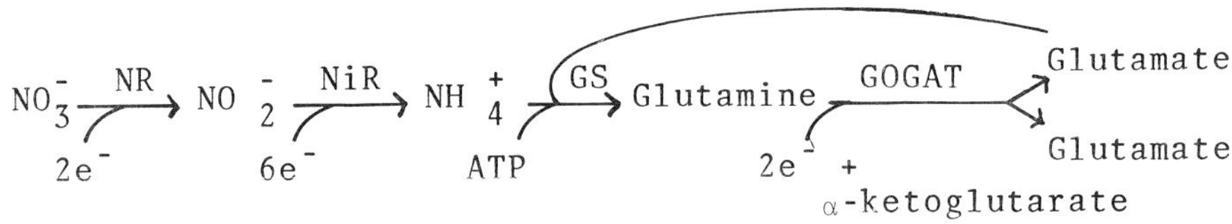

Nitrate Assimilation in Roots versus Shoots of Plants.

As noted by Lee (13), nitrate assimilation has been examined in non-photosynthetic tissues of several species with most emphasis on the roots. Plant species differ widely in the proportion of incoming nitrate that is reduced in the roots as compared to the shoots (6,47). Relative activities of NR in different plant parts vary with species as does the relative distribution of nitrate and reduced N in exuding xylem sap. It is not clear why differences in the relative proportions of nitrate reduced in roots exist among genotyes and species. Jackson (6) discussed the potential competition between reduction and xylem deposition components of the nitrate assimilation pathway in roots. He suggested that NR in the roots appears to have priority over xylem deposition for incoming nitrate. He reported that the rate of deposition of nitrate into the xylem differed significantly in three maize hybrids. These hybrids differed little in the amount of nitrate absorbed during the course of the experiment, but the amount of nitrate translocated out of the root system appeared to be inversely related to the amount of nitrate reduced in the roots.

In vitro NR activity has been assayed in the root tissue of a number of plant species (6,13). The activity is generally low compared with NR activity in leaves. Wallace (48) found that in maize seedlings, less than 20% of the total NR was found in the roots. Robin *et al.* (49) observed that NR activity in maize roots was only 20% of that in shoots. Solahuddin and Schrader (unpublished) recently compared *in vitro* root and shoot NR activity in 16 maize inbreds. Although casein was added to stabilize the NR activity (50), root NR activity per g fresh weight was generally only about 10% of the shoot NR activity in each genotype. Significant genotype differences in NR activity were observed in both root and shoot extracts. Maize genotypes with higher levels of shoot NR activity also had higher levels of root NR activity as indicated by a highly significant correlation (r=0.786**) when root and shoot NR activity per gram of tissue for the 16 genotypes were compared. When total NR per shoot was compared to total NR per root, the correlation was also highly significant (r=0.787**). Brunetti and Hageman (51) compared levels of extractable NR with the *in vivo* assay in leaves and roots of wheat plants, and found that although much higher levels of extractable NR were present in shoots versus roots, the *in vivo* assays indicated approximately equal levels of activity in the two tissues. Hence the low activities of NR observed in extracts from roots may need to be examined further due to the presence of an inactivating enzyme in root extracts (52-55).

The other enzymes involved in nitrate assimilation have not been studied as extensively in non-green tissues as has NR, but Lee (13) has recently reviewed the status of work in that area. Major differences may exist between green and non-green tissues in the processes that provide the reductant needed for nitrate assimilation.

Intracellular Localization of Enzymes.

With respect to subcellular localization of the enzymes of nitrate assimilation, most evidence suggests that NR is a soluble enzyme localized in the cytoplasm of both leaf and root tissue (13). A particle-bound NR was prepared by Miflin (56-58) from non-sterile barley roots, but Blevins et al. (59) later showed that the particle-bound form was probably of microbial origin. Butz and Jackson (17) proposed the existence of a membrane-bound enzyme with nitrate transporting, nitrate reducing, and ATPase activities. This proposal would account for the frequently observed correlation between nitrate absorption and nitrate reduction, but more evidence is needed before this proposal can be accepted. In leaves, Rathnam and Das (60) and Rathnam and Edwards (61) reported that NR is localized on the outer envelope of the chloroplast and thus is easily solubilized during extraction.

NiR is localized within the chloroplasts in leaf tissue (62-64). Evidence has been presented (65,66) that NiR in non-green tissue is associated with proplastids.

Lee (13) recently reviewed the evidence that GS is localized both within the plastids and cytosol in green and non-green tissues. Hence the enzyme appears to be present to detoxify ammonia formed by nitrite reduction in chloroplasts and proplastids, and ammonia released during photorespiration and other deamination reactions (67).

GOGAT is localized in the chloroplasts (64,68-70) and appears to be associated with proplastids in non-green tissues (44,66).

Support for the findings cited above is provided by studies (71,72) in which nitrite was reduced to ammonia and amino acids by isolated chloroplasts in a light-dependent, CO_2-independent reaction, without other additives. Anderson and Done (69,70) have shown that isolated chloroplasts catalyze nitrite-dependent and glutamine plus α-ketoglutarate-dependent oxygen evolution. These results indicate that nitrite reduction and glutamate synthesis are dependent on reductants generated by the chloroplast electron transport chain.

Intercellular Localization of Enzymes.
Rathnam and Edwards (61) studied the intercellular localization of nitrate-assimilating enzymes in three C_4 groups of plants. NR, NiR, GS and GOGAT were predominantly localized in mesophyll cells of the species studied. Harel et al. (73) found more GS and GOGAT in maize bundle sheath cells than did Rathnam and Edwards (61) and attributed the differences to improved extraction procedures. Neyra and Hageman (74) found 95% of the NR and NADP-malic dehydrogenase in mesophyll cells of maize. They reported that the nitrate concentration was 16 times higher in mesophyll cells than in bundle sheath cells, suggesting the direct transport of nitrate to mesophyll cells from the xylem (not through bundle sheath cells). Attempts to induce NR with nitrate in bundle sheath strands failed. NAD-malate dehydrogenase and NAD-glyceraldehyde-3-P dehydrogenase were about equally distributed between the two cell types, but NAD-malate dehydrogenase was about 10 times higher in activity than was glyceraldehyde-3-P dehydrogenase in mesophyll cells. On the basis of enzyme activities and the fact that malate is formed during CO_2 fixation in the mesophyll cells of C_4 species, it seems likely that malate oxidation is an important and perhaps the major source of NADH for NR in C_4 species. Glyceraldehyde-3-P is formed in bundle sheath chloroplasts by the Calvin cycle and must be transported from the bundle sheath cells to the mesophyll cells to provide reductant for reactions in the mesophyll cells. Moore and Black (75) reported similar

results for localization of these enzymes in crabgrass. Mesophyll protoplasts contained most of the NR and NiR, but GS was equally distributed between mesophyll and bundle sheath cells, and GOGAT was higher in bundle sheath cells. They suggest that nitrate assimilation in mesophyll cells results in a "division of labor" between mesophyll and bundle sheath cells, and that a spatial separation of the oxygen evolution accompanying nitrite reduction (in mesophyll cells) from carbon dioxide fixation (in bundle sheath cells) helps keep the oxygen away from the ribulose bisphosphate carboxylase/oxygenase. This may reduce the level of photorespiration.

Electron Donors for Nitrate Assimilation.

Nitrate reductase from a variety of green and non-green tissues has been shown to use NADH as the electron donor (13,76), although NR from some tissues has the ability to use NADPH as well (77-81). The NAD(P)H NR does not appear to be highly specific for NADPH, whereas NADH-NR from many tissues is specific in its requirement for NADH (13).

Ferredoxin, a natural one-electron carrier in plants, provides the reductant for NiR in illuminated chloroplasts (82). The reductant for non-green tissues has not been identified. Reduced pyridine nucleotides do not donate electrons directly to NiR although there is some evidence that a reductase exists to transfer electrons from pyridine nucleotides through ferredoxin to NiR (83). Because the properties of NiR from green and non-green tissues are so similar (84-86), it seems likely that a compound similar to ferredoxin also serves as the reductant in non-green tissue. However, ferredoxin has not yet been isolated from non-green plant tissue, but perhaps a one-electron carrier similar to ferredoxin exists in these tissues.

Ferredoxin also serves as the immediate electron donor for GOGAT purified from green leaves, but pyridine nucleotides are ineffective donors (64,87). In non-green tissues, GOGAT generally uses NADH as a reductant more effectively than NADPH (88,89).

Relationship of Nitrate Assimilation and Photosynthesis. Nitrate assimilation in green tissue is dependent on photochemical reactions of the chloroplast to provide the reductant and ATP used to convert nitrate to amino acids (Fig. 1). Nitrate reduction in the cytosol can be provided indirectly with reductant (NADH) from photosynthesis by oxidation of substrates (e.g., dihydroxyacetone phosphate [DHAP] or malate) that are transported from the chloroplasts to the cytosol by the phosphate translocator or dicarboxylate translocator (90,91). Inside the chloroplasts, the reduction of nitrite to ammonia can be coupled directly to light reactions through reduced ferredoxin as the electron donor. ATP from photophosphorylation can be used for the GS reaction and reduced ferredoxin can be used directly for the GOGAT reaction. As noted in Fig. 1, the quantum requirement for reduction of nitrite to ammonia is 12. The conversion of glutamine to two glutamates requires four photons. If a triose phosphate (e.g., DHAP) is translocated from the chloroplast to the cytosol to provide the NADH for nitrate reduction, the reduction of the PGA to glyceraldehyde-3-P in the chloroplast will require four photons (Fig. 1). Therefore in illuminated green tissues, the total quantum requirement for nitrate reduction to ammonia is 16 photons per nitrate ion, and for conversion of nitrate to glutamate is 20.

For non-chlorophyllous tissue or dark reduction of nitrate in green tissues, the reductant for nitrate assimilation must come from oxidation of carbohydrates or organic acids. On the assumption that most of the carbon is provided by sucrose transported to the site of reduction, each hexose can provide 12 pairs of electrons if totally oxidized to CO_2 and H_2O in glycolysis and the Krebs' tri-

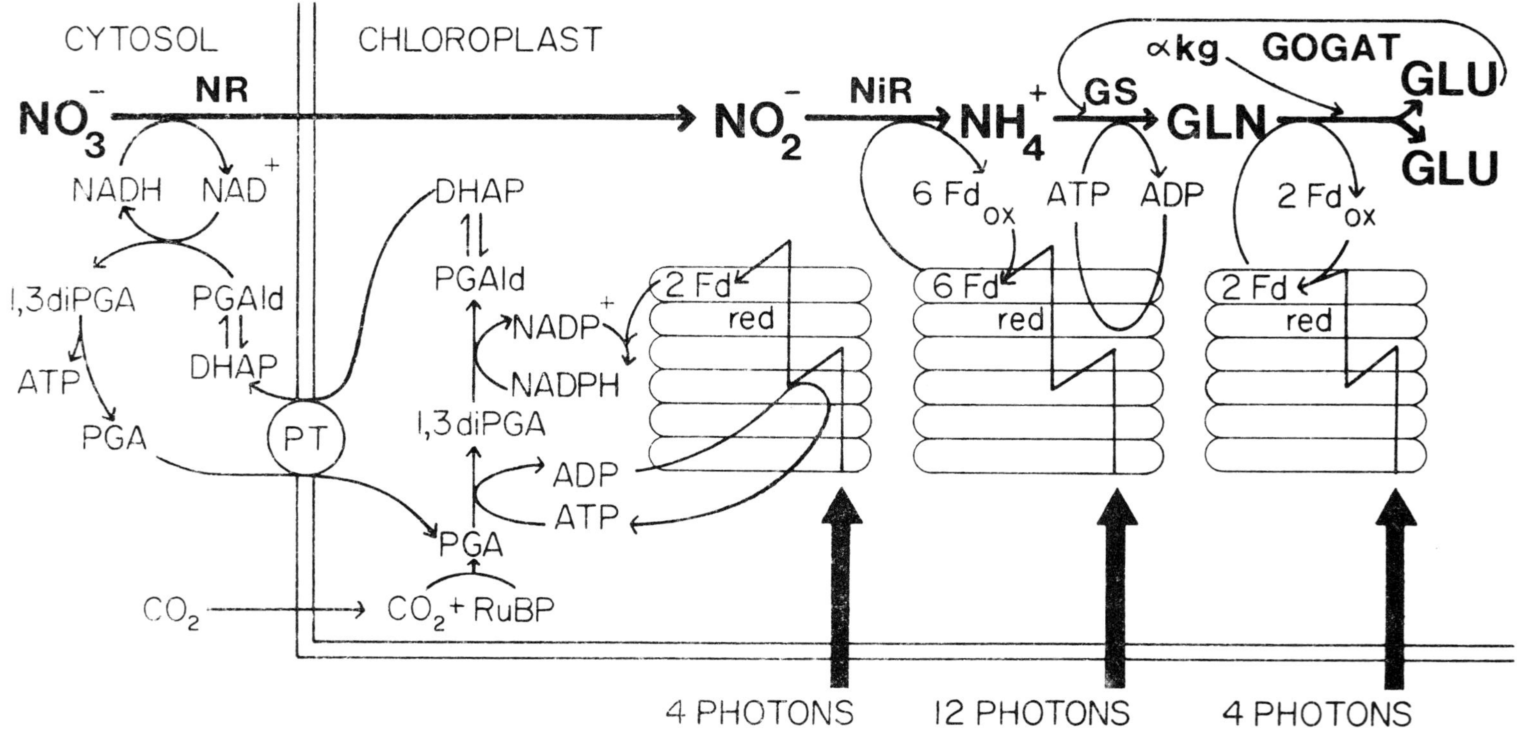

Fig. 1. *The relationships between nitrate assimilation and photochemical reactions in the chloroplast. PT = phosphate translocator; NR = Nitrate reductase; NiR = nitrite reductase; GS = glutamine synthetase; GOGAT = glutamate synthase; GLN = glutamine; GLU = glutamate;* Fd_{red} *= reduced ferredoxin;* Fd_{ox} *= oxidized ferredoxin; DHAP = dihydroxyacetone phosphate; PGAld = 3-phosphoglyceraldehyde; 1,3 diPGA = 1,3 bisphosphoglycerate; PGA = 3-phosphoglycerate; RuBP = ribulose, 1,5-bisphosphate; αkg = α-ketoglutarate*

carboxylic acid cycle (92). Therefore each carbon of the hexose can provide two pairs of electrons. To obtain the 10 electrons required for conversion of each nitrate ion to glutamate, the equivalent of 2.5 fixed carbons must be oxidized. If one accounts for the photorespiratory losses normally associated with CO_2 fixation, the quantum requirement for fixation of carbon in plants is at least 14 to 15 (93). Therefore the quantum requirement for nitrate reduction in non-green tissue or green tissue in the dark appears to be almost double that in green tissue in the light.

If excess reductant is present in chloroplasts under high-light conditions (94), nitrate reduction in green leaves in light may be essentially free to plants. As noted above, reduction of nitrate in the dark or in non-green tissue is expensive. Nitrate reduction in roots and other non-green tissues may be important in providing a continual supply of reduced nitrogen to other plant parts during the night. However, from an energetic point of view and from the standpoint of maximizing productivity, it seems desirable to reduce a large portion of the nitrate in the light in green tissues.

Regulation of Nitrate Reductase.

The regulation of NR is complex and appears to differ from species to species as well as in different plant parts. This topic has ben reviewed by several authors (6,7,13,40,42,95-97) and the interested reader is referred to these. The various regulatory factors can be divided into those that affect (a) the amount of enzyme and (b) the activity of existing enzyme.

a) Regulation of amount of enzyme

Nitrate has been shown to induce the de novo synthesis of NR (42,98). Light has also been shown to be involved in the induction process (95,99) although its requirement is not absolute (100). The effects of light on the formation of new enzyme are numerous and include (a) increasing the uptake of nitrate into leaves (101), (b) maintaining polyribosome levels and hence the capacity for protein synthesis (100), (c) acting via a phytochrome system which affects the movement of nitrate across membranes, e.g., from a "storage" pool to a "metabolic" pool (102,103), and (d) inactivating an inhibitor of NR which is formed in the dark (104).

Much of the work on the "induction" of NR, while showing that protein synthesis is necessary, does not distinguish between *de novo* synthesis of the enzyme and assembly of pre-existing polypeptides. Moreover, decreased rates of inactivation or degradation of NR could result in higher levels of extractable NR activity (97,105). There is some evidence that NR may be a constitutive enzyme (97). Ammonium-grown *Chlorella* cells with almost no NR activity were shown to contain cross-reacting material which binds to NR antibody suggesting the presence of a NR precursor protein (106). These results together with inhibitor and labelling studies (107,108) suggest that NR mRNA and precursor protein are present in cells which have not been subjected to inducing conditions (presence of nitrate). The so called "induction" of NR by nitrate may thus involve an assembly of pre-existing enzyme components and not strictly a *de novo* synthesis (see 97 for detailed discussion of enzyme component assembly). Further serological work is required to test this suggestion in higher plants which generally contain low NR activity when grown with ammonium.

b) Regulation of nitrate reductase activity

Numerous factors affect NR activity. The availability of substrate (nitrate) and reductant (usually NADH) are key regulatory factors, but other factors that inactivate or degrade the enzyme are involved in regulation too.

Perhaps the most important factor in the regulation of NR in leaf tissue is the flux of nitrate into the leaves from the roots or storage tissues. Following the work of Meeker *et al.* (109) who showed that leaf NR activity was more closely correlated with nitrate concentration of the midrib of maize leaves than with the concentration in the lamina, Shaner and Boyer (110) showed that the nitrate flux into the leaves was more important in the regulation of NR than was leaf nitrate content. The results can be interpreted in terms of there being two pools of nitrate in the leaf tissue: (1) a storage pool (possibly in the vacuole), and (b) a smaller rapidly metabolized pool in the cytoplasm (111). According to this proposal (111), NR is regulated by the flux of nitrate entering the metabolic pool via the transpiration stream. The storage pool would contribute largely to the total leaf nitrate content but little towards the metabolic pool because of sequestration in the vacuole (110). In this way, changes in NR activity can occur without significant changes in total leaf nitrate content.

Availability of NADH for NR in leaf tissue provides another regulatory system. The reduction of ^{15}N-nitrate and -nitrite in leaves has been shown to cease in the dark (112). However, nitrate reduction in the dark can be promoted by anaerobic conditions (113), or by inhibiting the respiratory electron transport chain with CO (114) or antimycin A, but not amytal or rotenone (113). This promotion of nitrate reduction in the dark has been attributed to availability of extramitochondrial NADH at times when the external NADH dehydrogenase of the mitochondria is inhibited (113). In the light, the respiratory electron transport chain is thought to be inhibited by a high ATP:ADP ratio that prevails in leaves (115). Further support for these proposals was provided recently by Woo *et al.* (116) with a reconstituted system from spinach leaves containing supernatant, mitochondria, NAD^+, oxaloacetate and an oxidizable substrate. They showed that the transfer of reducing power from the mitochondria to the soluble phase via the oxaloacetate/malate shuttle can provide NADH for the cytoplasmic reduction of nitrate. This rather elegant regulatory system in leaves serves at least two important roles: (a) it prevents the accumulation of nitrite in leaf tissue during the night when photochemical reductants (i.e., ferredoxin) are not being produced in the chloroplast, and (b) it requires the leaf to use light-dependent reductants rather than fixed carbon sources as a source of reductant. In contrast to the results cited above, Aslam et al. (117) reported that dark reduction of nitrate in barley seedlings was about 50% of that observed in the light. These conflicting results cannot be explained at this time. Roots and other non-photosynthetic tissues apparently depend on factors other than NADH availability for regulation of NR, as oxidative phosphorylation in mitochondria in those tissues presumably competes for NADH thoughout the photoperiod.

Light is thought to be involved in activation of the enzyme via several direct and indirect effects (42,95,97,118) including a direct activation of the enzyme (119), an increase in accessibility of the enzyme to nitrate via phytochrome-mediated membrane changes and/or other phytochrome effects (120), and provision of reductant via photosynthesis (121,122).

The enzyme can be converted from an active to an inactive form by various chemical changes including an "over-reduced" state with NAD(P)H or dithionite (123). These effects can be enhanced by the presence of cyanide (123). Further evidence that cyanide is involved in the regulation of NR in algae has been published (124)and a model linking the regulation of CO_2 fixation and nitrate assimilation and involving the production of cyanide has been proposed (118). The NAD(P)H- and cyanide-inactivated form of the enzyme can be reactivated by ferricyanide which is thought to reoxidize the enzyme (123). These results suggest that NR activity may be controlled by the redox state of the cells (125). Hewitt et al. (97) have obtained evidence that a similar regulatory effect of cyanide exists in higher plants but further work is required before this suggestion can be generally accepted as a major regulatory factor of NR *in vivo*. Reversible inactivation also occurs with ADP and other adenine nucleotides (97).

In green algae and simple aquatic plants such as *Lemna*, ammonium has been shown to repress or prevent the increase in extractable NR activity in the presence of nitrate (126-128). There is still controversy over whether or not the effect of ammonia is on NR itself or if it first inhibits nitrate uptake into algal cells (126). In most higher plants studied, ammonium does not repress NR (7,40,42). Furthermore although Orebamjo and Stewart (129) observed an inactivation of NR by addition of ammonium to nitrate-induced *Lemna*, similar experiments using *in vitro* assays of NR in maize revealed that monovalent cations (NH_4^+, Na^+ and K^+) at equal concentrations had similar inhibitory effects on NR (7). These results suggest that ionic strength rather than a specific effect of ammonium ions is the critical factor.

In general amino acids, as eventual products of nitrate reduction, do not inhibit leaf NR activity or induction (7,40,96). There is some evidence that amino acids may regulate the induction of NR in root tissue (130,131) and plant tissue cultures (132). These results suggest that there may be different regulatory mechanisms operating in the root and shoot.

NR-inactivating proteins have been isolated from corn roots (52), rice cell cultures (133) and seedlings (134), soybean leaves (104), wheat (135,136) and cucumber leaves (137). These proteins may be responsible for the loss of NR activity following removal of nitrate from the medium and for the frequently observed diurnal rhythms in NR activity (135). The proteins isolated from the different tissues studied appear to have different characteristics

(135,138). For example, the protein isolated from corn roots has proteolytic activity (52) whereas that isolated from rice plants and soybean leaves has no proteolytic activity (104,138,139). Thus there appear to be at least two types of protein, one which is a NR-degrading protein (52) and the other a NR-specific binding protein (140). In addition to these inactivating factors, there is some evidence from studies with wheat and soybean leaves for the presence of activating factors (104,135). Sherrard _et al_. (136) have suggested that because of the presence of these activating factors, it may be premature to describe the inactivating proteins as being specific for NR, as a negative result of an inactivating factor on any enzyme may be due to the presence of contaminating protective factors.

The effects of CO_2 on nitrate reduction are complex (7), but generally CO_2 in the presence of light enhances nitrate reduction probably by increasing the amount of fixed carbon available to provide reductant. Increasing concentrations of CO_2 may decrease leaf NR activity by decreasing stomatal opening and hence nitrate flux into the leaves (29,109).

The level of NR activity in plant tissues is dependent to some extent on other nutrient elements. Molybdenum is a constituent of the enzyme and must be present for synthesis of active NR (42). Because the enzyme is sulfhydryl-dependent (141), adequate sulfur must be present to induce and maintain high levels of NR activity (142). Friedrich and Schrader (142) observed that S-deprived maize seedlings lost NR activity more quickly than other enzymes involved in nitrate assimilation, and prior to any decrease in soluble protein. Furthermore, NR activity had decreased to 50% of the control before any visual symptoms of S deficiency began to appear. With stricter pollution control of our atmosphere, the removal of more SO_2 from the atmosphere may cause marginal S deficiency to become more prevalent in crops. It will be important to watch carefully for this condition, as nitrate assimilation may be severely limited by even marginal deficiencies of sulfur.

Temperature also influences NR activity (96). Increasing temperatures may increase NR activity for short periods of time, but prolonged periods of high temperature result in lower NR activity, although the magnitude of inactivation varies according to species (96). In some cases, lower than optimal temperatures afford some protection to the inactivation of NR activity (143). Onwueme et al. (143) reported a protective effect of low temperature against dark inactivation of leaf NR in barley. Alofe et al. (144) studied the effects of high day temperatures (35°C) and different night temperatures (7,16, and 24°C) on NR activity of 16 genotypes of maize under controlled environment conditions. They observed that NR activity was much lower at the high night temperature. Deane-Drummond et al. (25) observed that both in vitro and in vivo NR activity in roots of barley were higher when roots were grown at 10°C with the shoot at 20°C compared with whole plants grown at 20°C. This apparent adaptive response may be relevant to field situations in which soil temperatures are generally lower than air temperatures. The contribution of the root system to nitrate assimilation of the whole plant may need to be reassessed.

In many cases, high temperature stresses may be associated with water stress (145), so that it is difficult to ascertain whether temperature or water stress is responsible for the observed decreases in NR activity. Decreases in water potential below -4 to -2 bars causes NR activity to decrease in barley (145) and maize (147,148).

Relationship of NR to Grain Protein and Yield.

Because NR is thought to be the rate-limiting enzyme in nitrate assimilation (76), Hageman and others have studied the relationship between NR activity and grain yield and protein in cereals. Hageman (146) recently reviewed several experiments conducted in his laboratory (149-152) in which NR activity was shown to be correlated with either yield or protein in wheat or maize. In other laboratories (153-157) a positive relationship has also been demonstrated. Less successful results have been obtained by others (158-160).

It is not surprising that some of the results have been variable when one considers the complexity of the regulation of NR. Several environmental factors, nitrate availability, and the supply of reductant (NADH) in leaf tissue are some of the factors which participate in the complex regulation of NR in leaf tissue. A highly positive correlation between NR activity and yield will not likely be observed except in those environments in which nitrogen is the factor most limiting yield.

TRANSPORT OF REDUCED NITROGEN

Much of the reduced N in a plant must be transported to other cells at least once during the life cycle of a plant. As we noted earlier, nitrate may be reduced in roots or other non-green tissues and in leaves. The site of nitrate reduction influences the transport of reduced N. Although the xylem is the principal path for long-distance transport of nitrogenous solutes and roots to organs that transpire (47,161-164), there are reported exceptions. In *Phaseolus vulgaris* upward transport of ^{15}N has been reported in the xylem and phloem (165). Joy and Antcliff (166) found that readily soluble organic N solutes may be transported upward in the phloem of *Beta vulgaris*. The principal path for export of reduced N from leaves is the phloem (161,162,167,168).

Leaves and other vegetative tissues do not export all the reduced N produced during nitrate reduction. Instead they store considerable amounts of proteins (169) which become important sources of reduced N during later stages of grain-fill in many species (169-173). Remobilization of these proteins becomes important because nitrate reduction is usually insufficient to meet the demands of the developing seeds for reduced N during this period (146). One of the reasons for this deficit of reduced N is the inadequate supply of nitrate in leaves to provide saturating levels of substrate (nitrate) for leaf NR (149,152).

Composition of Xylem and Phloem Saps in Plants Grown on Nitrate.

According to Pate (162,174), N solutes frequently comprise the major component of dry matter in xylem sap, and are second only to carbohydrates in the phloem sap. The C:N weight ratio in xylem sap ranges from 1.5 to 6, whereas the ratio in phloem sap ranges from 15 to 200.

The composition of xylem sap varies considerably from species to species (47,162,174). The ratio of organic N:nitrate-N also varies widely because species differ in their ability to assimilate nitrate in the roots (47,162,164). Over 95% of the xylem N may consist of nitrate in some species that have low NR activity in their roots (e.g., Cucumis, Xanthium and Gossypium) whereas other species with high root NR activity (e.g., Pisum, Raphanus and Lupinus) have less than 20% of the xylem N as nitrate-N (47,162). Reduced N in roots is incorporated into a limited number of amino acids, amides and other solutes for transport to the shoot, but each plant species seems to have a characteristic spectrum of these N compounds (47,164,174). The most common N solutes in plants grown on nitrate are aspartate and glutamate and their amides (47,164,174). Ureides and other solutes play an important role in xylem transport in certain species grown on nitrate (175), but ureides are most important as a transport solute in certain tropical legumes that are dependent on nitrogen fixation for much of their reduced N

(176,177). Ureide assimilation will therefore not be discussed here.

The types of N solutes found in the phloem sap also vary from species to species, but because of the difficulties of collecting phloem sap from many species, the contents of only a limited number of species have been examined (162,174). In contrast to the xylem, N solutes in the phloem are organic solutes (162,174). Nitrate is usually absent from, or present in only trace amounts in the phloem (161,178,179). Asparagine was highest in the phloem of the legume Lupinus (179). Frequently the glutamine/glutamate and asparagine/aspartate fractions are both high, as observed in Ricinus phloem exudate (180). Serine, however, was found to be the amino acid exported from young soybean leaves in the greatest quantity (181). Later studies showed that aspartate, glutamate, alanine, and gamma-aminobutyric acid were also transported in significant amounts in soybean petioles (182,183). Amides also appear in the phloem sap of some species at the time of senescence (184,185) and are considered to result from ammonium production often accompanying protein hydrolysis (186).

Di- and oligopeptides comprise an additional class of nitrogenous compounds that may be important in transport. Duke et al. (187) reported large quantities of peptides in the axes of developing soybean seedlings, suggesting that peptide transport may be occurring. Higgins and Payne (188,189) have demonstrated active transport of dipeptides and oligopeptides up to five amino acids in length into the scutellum of germinating barley embryos. They suggest that in view of the energetics of membrane transport, a plant can conceivably transport a smaller number of peptide molecules more efficiently than an equivalent quantity of free amino acids. If the phloem of leaf tissue is also capable of loading and translocating peptides, these compounds may be an important form of

N transport in the mature plant. This mechanism may be particularly important during senescence when high rates of protein hydrolysis occur in the leaves.

Phloem Loading and Transport of Nitrogenous Solutes. Housley et al. (168) verified that organic N is exported via the phloem from soybean leaves. A petiole was heat girdled, disrupting phloem but not xylem functioning. The attached leaf was exposed to $^{14}CO_2$ for 2 hours, but no radioactive sugars or amino acids moved out of the fed leaf, thus demonstrating the importance of the phloem in exporting amino-N from leaves. These results were consistent with earlier studies (190,191).

Subsequent studies of amino acid export from soybean leaves followed the transport and conversions of photosynthetically-derived ^{14}C-amino acids (182). Only five of the amino acids in the transport pathway contained substantial amounts of ^{14}C after a 2-hour exposure of a source leaf to $^{14}CO_2$. The distribution of ^{14}C in aspartate and gamma-aminobutyrate was much higher in the path than in the source leaf, whereas serine, alanine and glutamate were substantially labeled in both the source leaf and the path. The total percentage of recovered ^{14}C in amino acids also differed, with 8 to 17% in leaf amino acids compared to only 2 to 6% in the petiole amino acids. These findings suggest some selectivity in the efflux processes from mesophyll cells into the apoplast, and/or in the loading processes. In contrast, Housley et al. (168) applied several amino acids individually to an abraded spot on a mature leaf, and found that amino acids not typically occurring in the phloem (lysine and leucine) were loaded as readily as those such as serine which are common in the phloem. Transport velocities for these amino acids also were similar to those obtained for ^{14}C-photosynthates produced by feeding a similar leaf with $^{14}CO_2$ (168). Servaites et al. (192) found that the rates of transport of ^{14}C-amino acids applied to an abraded spot on the leaf were similar to those of sucrose, and were not dependent on the nature of the functional group of the amino acid (i.e., leucine [neutral] vs. glutamate [dicarboxylic] vs. gamma-aminobutyrate [lacking alpha-amino group]). Sucrose neither inhibited nor enhanced the loading of leucine when applied with leucine to the abraded spot (192) suggesting separate and independent carriers. On the basis of these studies, the observed selectivity noted earlier (182) could be effected in the process by which amino acids are transferred from the mesophyll cells into the apoplast for loading.

Little is known about the mechanism of phloem loading of amino acids. For example, what is the phloem's capacity for transporting N solutes? Is active transport involved? Are specific carriers involved in loading each amino acid, or does a less specific carrier exist for a group of related amino acids?

The possible involvement of carriers for amino acid uptake has been implicated in studies of tissues and cells other than those of the phloem. Amar and Reinhold (193) reported that amino acid uptake by bean leaves was severely reduced by osmotic shock or EDTA, suggesting that a protein involved in transport was released from the membranes. Absorption studies also suggested the existence of amino acid carriers since uptake rates could be saturated with increased concentrations in the surrounding medium (194). This uptake pattern had two saturation phases, as did those of five other amino acids in similar experiments using cultured tobacco cells (195). These biphasic responses indicate multiple carriers, which were examined by Berlin and Mutert (195). Three separate carriers, for neutral, basic and acidic amino acids, were proposed on the basis of competition studies. Further evidence indicated that neutral amino acid loading took place via all three carriers, that acidic amino acids could be loaded on both the acidic and basic carriers, and that only basic amino acids were specifically transported by their own carrier (195). This theory is also supported by earlier work (196) on cultured soybean root cells. Similar competition studies, again between neutral, basic, and acidic amino acids, suggested at least three amino acid uptake carriers. The properties of such amino acid carriers have been examined in non-plant tissues (197), but evidence for such systems is scarce in plants, and especially for the phloem.

The involvement of a carrier, or carriers, in phloem loading of amino acids and sucrose in soybean leaves was suggested by the concentration dependence of sucrose and leucine loading and transport (192). The amount of leucine or sucrose loaded and transported increased as the concentration applied to the abraded spot was increased, until saturation levels were reached. The uptake of ^{14}C-sucrose from the free space of an abraded spot followed a biphasic saturation pattern and was similar to that observed in several other tissues (198). Although the amino acid leucine also showed saturation responses typical of carrier-mediated transport, the uptake pattern was triphasic rather than biphasic (192). The apparent K_m's for the leucine uptake phases were 3, 21, and 52 mM, respectively, compared to apparent K_m's of 35 and 100 mM for sucrose. The existence of distinct carrier systems for leucine and sucrose was further demonstrated by competition studies. When ^{3}H-leucine and ^{14}C-sucrose were supplied in a 1:4 concentration ratio to an abraded spot, the transport characteristics of leucine remained unchanged. Sucrose at 100 mM (about five times more concentrated than would be expected in the apoplast) also had no effect on the loading of 10 mM leucine (192). The significance of the three phases for leucine absorption must yet be determined, but they do indicate different carrier systems for sucrose and amino acid loading.

Furthermore, these data suggest that the carrier for leucine has a higher affinity than the one for sucrose.

In soybean leaves, both amino acid and sucrose loading were inhibited by the presence of the uncouplers DNP (2,4, dinitrophenol) and CCCP (carbonylcyanide-M-chlorophenyl hydrazone) (192). This suggested that phloem loading or transport of amino acids is metabolically dependent. The involvement of sulfhydryl groups in phloem loading of amino acids was also shown by using PCMBS (p-chloromercuribenzene sulfonic acid) which complexes with sulfhydryl groups (199). Applications of high concentrations of KCl inhibited both amino acid and sucrose loading by approximately 50% in soybeans. These effects of potassium concentration and the inhibition of amino acid loading by uncouplers are consistent with the model proposed for K^+- and H^+-mediated transfer of sucrose (200) and amino acids (201). According to this model, solute movement into the phloem is mediated by specific carriers and occurs with simultaneous co-transport of a proton which partially neutralizes the proton gradient across the sieve tube plasmalemma (200). To re-establish the protein gradient, ATP is hydrolyzed by a plasmalemma ATPase and protons are extruded from the sieve tube into the free space. High concentrations of KCl and uncouplers could dissipate the proton gradient, and thus inhibit loading.

Transport of Nitrogen as a Possible Limiting Factor for Productivity.

In some crop species, such as soybeans, the demand for reduced N in developing seeds is extremely high (202). Soybean seeds contain 6 to 8% N at maturity. At 7% N in the seeds, a seed yield of 3,360 kg hectare^{-1} (50 bushels acre^{-1}) requires 235 kg N hectare^{-1} (210 lb N acre^{-1}) for seed alone. If the nitrogen harvest index is 0.67, 353 kg N hectare^{-1} (315 lb N acre^{-1}) are required for the production of aboveground (seed + stover) plant parts. Most of the N in the seed is in protein. Because protein is about 16% N, protein constitutes from 37.5 to 50% of the dry weight of seeds containing 6 to 8% N, respectively. To meet the heavy demand for N during seed development, Sinclair and deWit (203) estimated that up to 9 kg N $hectare^{-1}$ day^{-1} must be transported to the seeds. This demand exceeds the capacity of nitrogen fixation to provide reduced N, and may exceed the capacity of the plant's transport system to the seeds. Thus more information is needed on the properties of the carriers and mechanisms by which amino acids and other nitrogenous assimilates are transported to the seeds.

Although sucrose is the predominant solute in the phloem of most plant species, the following calculations will emphasize the importance of N transport during seed development of crops such as soybeans. Penning deVries (204) estimated that 1 gram of photosynthate (glucose-equivalent) can produce 0.83 g of carbohydrate, 0.33 g of lipid, 0.46 g of lignin, or 0.62 g of nitrogenous compounds (when ammonia-N is provided). Less protein can be produced per unit of photosynthate when nitrate is provided, but the value for ammonia-N will be used here because reduced N is transported to seeds. On the basis of these estimates, 1.74 grams of photosynthate will be required to produce a gram of soybean seed of the composition shown (Table 1). On the assumption that sucrose provides the carbon for energy production, and for carbohydrate, lignin, and lipid synthesis; and that amino acids provide the carbon skeletons for protein synthesis, amino acid transport would provide 0.42 g of dry matter (42% protein by dry weight) and sucrose transport would provide the balance (1.74 g minus 0.42 g = 1.32 g sucrose) of the photosynthate used for production of dry matter and energy. Therefore about three times as much carbon would be provided by sucrose as compared to amino acids. However, sucrose has 12 carbons per molecule whereas the

Table 1. Photosynthate requirements for production of soybean seeds.

Composition of soybean seed*	g photosynthate/g of seed
42% protein	0.677
20% lipid	0.606
31% carbohydrate	0.373
4% lignin	0.087
3% ash	----
100%	1.743

* Based on values used in (203).

average amino acid has about four carbon atoms. Hence one would expect one amino acid molecule for every sucrose molecule in the phloem during seed development. This would provide for a C:N ratio of 16:1 in the phloem of the transport pathway between vegetative tissues and developing fruits. These estimates would be modified somewhat for those species in which amides or ureides play an important role in transporting reduced N to the seeds.

Based on these assumptions and calculations, it seems conceivable that transport of reduced N to seeds could limit productivity. The frequently observed failure of foliar-applied fertilizers to increase soybean yields (205,206) could be due to a limited capacity of the plants to transport N to the seeds from leaves and other vegetative tissues. The concentration of N, P, and K increased in response to foliar applications of these elements (206,207), but yields were not increased (206).

SUMMARY

Nitrate uptake, assimilation and transport are extremely important processses for most crop species. Nitrogen frequently limits productivity, and therefore any improvements in the efficiency with which nitrate is absorbed and assimilated may result in higher protein and/or higher yields of these crops. These processes are complex and there are several control points at which regulation could be effected: nitrate uptake; transport of nitrate; accumulation of nitrate; remobilization of stored nitrate; reduction of the nitrate; transport of reduced N; and remobilization of stored N. Many, if not all, of these processes or phenomena are under genetic control. Hence one should be able to select for variability for these traits and then incorporate the desirable traits into new or existing genotypes to produce genotypes that are more efficient in utilization of nitrate-N.

REFERENCES

1. Allison, FE, Evaluation of incoming and outgoing processes that affect soil nitrogen. In Soil Nitrogen, Bartholomew, WV and FE Clark (eds.), Madison, WI, Amer. Soc. Agron, 573-606, 1965
2. Allison, FE, The fate of nitrogen applied to soils. Advan. Agron. 18:219-258, 1966
3. Allison, FE, The enigma of soil nitrogen balance sheets. Advan. Agron. 22:213-250, 1970
4. Viets, FG, Jr. and RH Hageman, Factors affecting the accumulation of nitrate in soil, water and plants. USDA Agric Handbook no. 413, 63 p., 1971
5. Huffaker, RC and DW Rains, Factors influencing nitrate acquisition by plants; assimilation and fate of reduced nitrogen. In: Nitrogen in the Environment, vol. 2, Nielsen, DR and JG MacDonald (eds.), London, Academic Press, 1-43, 1978
6. Jackson, WA, Nitrate acquisition and assimilation by higher plants: processes in the root system. In: Nitrogen in the Environment, vol. 2., Nielsen, DR and JG MacDonald (eds.), London, Academic Press, 45-88, 1978
7. Schrader, LE, Uptake, accumulation, assimilation, and transport of nitrogen in higher plants. In: Nitrogen in the Environment, vol. 2., Nielsen, DR and JG MacDonald (eds.), London, Academic Press, 101-141, 1978
8. Chantarotwong, W, RC Huffaker, BL Miller and RC Granstedt, *In vivo* nitrate reduction in relation to nitrate uptake, nitrate content, and *in vitro* nitrate reductase activity in intact barley seedlings. Plant Physiol 57:519-522, 1976
9. Friedrich, JW, LE Schrader and EV Nordheim, N depri-

vation in maize during grain-filling. I. Accumulation of dry matter, nitrate-N, and sulfate-S. Agron J 71:461-465, 1979
10. Friedrich, JW and LE Schrader, N deprivation in maize during grain-filling. II. Remobilization of ^{15}N and ^{35}S and the relationship between N and S accumulation. Agron J 71:466-472, 1979
11. Wright, MO and KL Davison, Nitrate accumulation in crops and nitrate poisoning in animals. Advan Agron 16:197-247, 1964
12. Beevers, L and RH Hageman, Nitrate reduction in higher plants. Annu Rev Plant Physiol 20:495-522, 1969
13. Lee, RB, Sources of reductant for nitrate assimilation in non-photosynthetic tissue: a review. Plant, Cell and Environ 3:65-90, 1980
14. Givan, CV, Metabolic detoxification of ammonia in tissues of higher plants. Phytochemistry 18:375-382, 1979
15. Miflin, BJ, SWJ Bright, HM Davies, PR Shewry and PJ Lea, Amino acids derived from aspartate; their biosynthesis and its regulation in plants. In: Nitrogen Assimilation of Plants, Hewitt, EJ and CV Cutting (eds.), London, Academic Press, 335-358, 1979
16. Goodman PJ, Genetic control of inorganic nitrogen assimilation of crop plants. In: Nitrogen Assimilation of Plants, Hewitt, EJ and CV Cutting (eds.), London, Academic Press, 165-176, 1979
17. Butz, RG and WA Jackson, A mechanism for nitrate transport and reduction. Phytochemistry 16:409-417, 1977
18. Haynes, RJ and KM Goh, Ammonium and nitrate nutrition of plants. Biol Rev 53:465-510, 1978
19. Clement, CR, LHP Jones and MJ Hopper, Uptake of nitrogen from flowing nutrient solution: Effect of terminated and intermittent nitrate supplies. In: Nitrogen Assimilation of Plants, EJ Hewitt and CV Cutting (eds.), London, Academic Press, 123-133, 1979
20. Rao, KP and DW Rains, Nitrate absorption by barley. I. Kinetics and energetics. Plant Physiol 57:55-58, 1976
21. Ben Zioni, A, Y Vaadia and SH Lips, Correlations between nitrate reduction, protein synthesis and malate accumulation. Physiol Plant 23:1039-1047, 1970
22. Ben Zioni, A, Y Vaadia and SH Lips, Nitrate uptake by roots as regulated by nitrate reduction products of the shoot. Physiol Plant 24:288-290, 1971
23. Frost, WB, DG Blevins and NM Barnett, Cation pretreatment effects on nitrate uptake, xylem exudate, and malate levels in wheat seedlings. Plant Physiol 61:323-326, 1978

24. Kirkby, EA and MJ Armstrong, Nitrate uptake by roots as regulated by nitrate assimilation in the shoot of castor oil plants. Plant Physiol 65:286-290, 1980
25. Deane-Drummond, CE, DT Clarkson and CB Johnson, The effect of differential root and shoot temperature on the nitrate reductase activity assayed _in vivo_ and _in vitro_ in roots of _Hordeum vulgare_ (barley). _Planta_, in press, 1980
26. Clement, CR, MJ Hopper, LHP Jones and EL Leafe, The uptake of nitrate by _Lolium perenne_ from flowing nutrient solution. _II. Effect of_ light, defoliation and relationship to CO_2 flux. J Exp Bot 29:1173-1183, 1978
27. Raven, JA and FA Smith, Nitrogen assimilation and transport in vascular land plants in relation to intracellular pH regulation. New Phytol 76:415-431, 1976
28. Kirkby, EA, and AH Knight, Influence of the level of nitrate nutrition on ion uptake and assimilation, organic acid accumulation and cation-anion balance in whole tomato plants. Plant Physiol 60:349-353, 1977
29. Neyra, CA and RH Hageman, Relationships between carbon dioxide, malate, and nitrate accumulation and reduction in corn (_Zea mays_ L.) seedlings. Plant Physiol 58:726-730, 1976
30. Minotti, PJ and WA Jackson, Nitrate reduction in the roots and shoots of wheat seedlings. Planta 95:36-44, 1970
31. Doddema, H and H Otten, Uptake of nitrate by mutants of _Arabidopsis thaliana_ disturbed in uptake or _reduction of nitrate._ III. Regulation. Physiol Plant 45:339-346, 1979
32. Smith, FA, The internal control of nitrate uptake into excised barley roots with differing salt content. New Phytol 72:769-782, 1973
33. Doddema, H and GP Telkamp, Uptake of nitrate by mutants of _Arabidopsis thaliana_ disturbed in uptake or _reduction of nitrate._ II. Kinetics. Physiol Plant 45:332-338, 1979
34. Epstein, E, Kinetics of ion transport and the carrier concept. In: _Transport in Plants_, Vol 2, Part B, Pirson, A and MH Zimmermann (eds.), Berlin and New York, Springer-Verlag, 70-94, 1976
35. Per Nissen, Uptake mechanisms: Inorganic and Organic, Annu Rev Plant Physiol 25:53-79, 1974
36. Clement, CR, MJ Hopper and LHP Jones, The uptake of nitrate by _Lolium perenne_ from flowing nutrient solution. _I. Effect of nitrate_ concentration. J Exp Bot. 29:453-464, 1978
37. Chevalier, P and LE Schrader, Genotypic differences in nitrate absorption and partitioning of N among plant parts in maize. Crop Sci. 17:897-901, 1977

38. Warncke, DD and SA Barber, Nitrate uptake differences of four plant species. J Environ Qual 3:28-30, 1974
39. Carpenter, EJ and RRL Guillard, Intraspecific differences in nitrate half-saturation constants for three species of marine phytoplankton. Ecology 52:183-185, 1971
40. Miflin, BJ, Nitrogen metabolism and amino acid biosynthesis in crop plants, In: The Biology of Crop Productivity, Carlson, PS (ed.), New York, Academic Press, 255-296, 1980
41. Haroon, M. Absorption, transport, accumulation, and remobilization of nitrogen in corn (Zea mays L.) at several stages of development. Ph.D thesis, Univ. of Wisconsin-Madison. Diss Abstr 36:3318B. 147p., 1975
42. Hewitt, EJ, Assimilatory nitrate-nitrite reduction. Annu Rev Plant Physiol 26:73-100, 1975
43. Miflin, BJ and PJ Lea, The pathway of nitrogen assimilation in plants. Phytochemistry 15:873-885, 1976
44. Miflin, BJ and PJ Lea, Amino acid metabolism. Annu Rev Plant Physiol 28:299-329, 1977
45. Hewitt, EJ and CV Cutting, Nitrogen Asssimilation of Plants (Sixth Long Ashton Symposium 1977), London, Academic Press, 708 p., 1979
46. Vennesland, B and MG Guerrero, Reduction of nitrate and nitrite. In Encyclopedia of Plant Physiol 6, Photosynthesis II, Gibbs, M and E Latzko (eds.), Berlin, Springer-Verlag, 6:425-444, 1979
47. Pate, JS, Uptake, assimilation and transport of nitrogen compounds by plants. Soil Biol Biochem 5:109-119, 1973
48. Wallace, W, The distribution and characteristics of nitrate reductase and glutamate dehydrogenase in the maize seedlings. Plant Physiol 52:191-195, 1973
49. Robin, P, D Blayac and L Salsac, Influence of l'alimentation nitrique sur la teneur en nitrate et l'activite nitrate reductase des racines et des feuilles de plantules de Mais. Physiol Veg 17:55-66, 1979
50. Schrader, LE, DA Cataldo and DM Peterson, Use of protein in extraction and stabilization of nitrate reductase. Plant Physiol 53:688-690, 1974
51. Brunetti, N and RH Hageman, Comparison of in vivo and in vitro assays of nitrate reductase in wheat (Triticum aestivum L.) seedlings. Plant Physiol 58:583-587, 1976
52. Wallace, W, A nitrate reductase inactivating enzyme from the maize root. Plant Physiol 52:197-201, 1973
53. Wallace, W, Purification and properties of a nitrate reductase inactivating enzyme. Biochim Biophys Acta 341:265-276, 1974

54. Wallace, W, A re-evaluation of the nitrate reductase content of the maize root. Plant Physiol 55:774-777, 1975
55. Yamaya, T, A Oaks and IL Boesel, Characteristics of nitrate reductase-inactivating proteins obtained from corn roots and rice cell cultures. Plant Physiol 65:141-145, 1980
56. Miflin, BJ, Distribution of nitrate and nitrite reductase in barley. Nature 214:1133-1134, 1967
57. Miflin, BJ, Studies on the sub-cellular location of particulate nitrate and nitrite reductase, glutamic dehydrogenase and other enzymes in barley roots. Planta 93: 160-170, 1970
58. Miflin, BJ, The location of nitrite reductase and other enzymes related to amino acid biosynthesis in the plastids of root and leaves. Plant Physiol 54:550-555, 1974
59. Blevins, DG, RH Lowe and I Staples, Nitrate reductase in barley roots under sterile, low oxygen conditions. Plant Physiol 57:458-459, 1976
60. Rathnam, CKM and VSR Das, Nitrate metabolism in relation to the aspartate-type C-4 pathway of photosynthesis in <u>Eleusine coracana</u>. Can J Bot 52:2599-2605, 1974
61. Rathnam, CKM and GE Edwards, Distribution of nitrate-assimilating enzymes between mesophyll protoplasts and bundle sheath cells in leaves of three groups of C_4 plants. Plant Physiol 57:881-885, 1976
62. Ritenour, GL, KW Joy, J Bunning and RH Hageman, Intracellular localization of nitrate reductase, nitrite reductase and glutamic acid dehydrogenase in green leaf tissue. Plant Physiol 42:233-237, 1967
63. Dalling, MJ, NE Tolbert and RH Hageman, Intracellular location of nitrate reductase and nitrite reductase. I. Spinach and tobacco leaves. Biochim Biophys Acta 283:505-512, 1972
64. Wallsgrove, RM, PJ Lea and BJ Miflin, The distribution of the enzymes of nitrogen assimilation within the pea leaf cell. Plant Physiol 63:232-236, 1979
65. Dalling, MJ, NE Tolbert and RH Hageman, Intracellular location of nitrate reductase and nitrite reductase. II. Wheat roots. Biochim Biophys Acta 283:513-519, 1972
66. Emes, MJ and MW Fowler, The intracellular location of the enzymes of nitrate assimilation in the apices of seedling pea roots. Planta 144:249-253, 1979
67. Keys, AJ, IF Bird, MJ Cornelius, PJ Lea, RM Wallsgrove and BJ Miflin, Photorespiratory cycle. Nature 275:741-743, 1978

68. Lea, PJ and BJ Miflin, Alternative route for nitrogen assimilation in higher plants. Nature 251:614-616, 1974
69. Anderson, JW and J Done, A polarographic study of glutamate synthase activity in isolated chloroplasts. Plant Physiol 60:354-359, 1977
70. Anderson, JW and J Done, Light-dependent assimilation of nitrite by isolated pea. Plant Physiol 61:692-697, 1978
71. Magalhaes, AC, CA Neyra and RH Hageman, Nitrite assimilation and amino nitrogen synthesis in isolated spinach chloroplasts. Plant Physiol 53:411-415, 1974
72. Miflin, BJ, Nitrite reduction in leaves; studies on isolated chloroplasts. Planta 116:187-196, 1974
73. Harel, E, PJ Lea and BJ Miflin, The localization of enzymes of nitrogen assimilation in maize leaves and their activities during greening. Planta 134:195-200, 1977
74. Neyra, CA and RH Hageman, Pathway for nitrate assimilation in corn (Zea mays L.) leaves. Cellular distribution of enzymes and energy sources for nitrate reduction. Plant Physiol 62:618-621, 1978
75. Moore, R and CC Black, Jr., Nitrogen assimilation pathways in leaf mesophyll and bundle sheath cells of C_4 photosynthesis plants formulated from comparative studies with Digitaria sanguinalis (L.) Scop. Plant Physiol 64:309-313, 1979
76. Beevers, L and RH Hageman. Nitrate reduction in higher plants. Annu Rev Plant Physiol 20:495-522, 1969
77. Elsner, JE, DP Hucklesby and RH Hageman, A nitrate reductase from corn scutella which can utilize NADPH as electron donor. Agron Abstr C2:20-21, 1971
78. Campbell, WH, Separation of soybean leaf nitrate reductases by affinity chromotography. Plant Sci Lett 4:249-256, 1976
79. Jolly, SO, W Campbell and NE Tolbert, NADPH- and NADH-nitrate reductase from soybean leaves. Arch Biochem Biophys 174:431-439, 1976
80. Shen, TC, EA Funkhouser and MG Guerrero, NADH- and NAD(P)H-nitrate reductases in rice seedlings. Plant Physiol 58:292-294, 1976
81. Campbell, WH, Isolation of NAD(P)H:nitrate reductase from the scutellum of maize. Z. Pflanzenphysiol 88:357-361, 1978
82. Joy, KW and RH Hageman, The purification and properties of nitrite reductase from higher plants, and its dependence on ferredoxin. Biochem J 100:263-273, 1966
83. Hucklesby, DP, MJ Dalling and RH Hageman, Some properties of two forms of nitrite reductase from corn (Zea mays L.) scutellum. Planta 104:220-233, 1972

84. Bourne, WF and BJ Miflin, Studies on nitrite reductase in barley. Planta 111:47-56, 1973
85. Dalling, MJ, DP Hucklesby and RH Hageman, A comparison of nitrite reductase enzymes from green leaves, scutella and roots of corn (Zea mays L.). Plant Physiol 51:481-484, 1973
86. Miflin, BJ, The location of nitrite reductase and other enzymes related to amino acid biosynthesis in the plastids of root and leaves. Plant Physiol 54:550-555, 1974
87. Rhodes, D, GA Rendon, and GR Stewart, The regulation of ammonia assimilating enzymes in Lemna minor. Planta 129:203-210, 1976
88. Beevers, L and R Storey, Glutamate synthetase in developing cotyledons of Pisum sativum. Plant Physiol 57:862-866, 1976
89. Chiu, JY and PD Shargool, Importance of glutamate synthase in glutamate synthesis by soybean cell suspension cultures. Plant Physiol 63:409-415, 1979
90. Heber, U, Metabolite exchange between chloroplasts and cytoplasm. Annu Rev Plant Physiol 25:393-421, 1974
91. Fliege, R, U Flügge, K Werdan and HW Heldt, Specific transport of inorganic phosphate, 3-phosphoglycerate and triosephosphates across the inner membrane of the envelope in spinach chloroplasts. Biochim Biophys Acta 502:232-247, 1978
92. Lehninger, AL, Biochemistry, New York, Worth Publishers, 833 p., 1971
93. Campbell, WH and CC Black, The relationship of CO_2 assimilation pathways and photorespiration to the physiological quantum requirement of green plant photosynthesis. Biosystems 10:253-264, 1978
94. Radmer, RJ and B Kok, Photoreduction of O_2 primes and replaces CO_2 assimilation. Plant Physiol 58:336-340, 1976
95. Beevers, L and RH Hageman, The role of light in nitrate metabolism in higher plants. Photophysiology 7:85-114, 1972
96. Srivastava, HS, Regulation of nitrate reductase activity in higher plants. Phytochemistry 19:725-733, 1980
97. Hewitt, EJ, DP Hucklesby, AF Mann, BA Notton and GJ Rucklidge, Regulation of nitrate assimilation in plants. In: Nitrogen Assimilation of Plants. Hewitt, EJ and CV Cutting (eds.), London, Academic Press, 255-287, 1979
98. Zielke, HR and P Filner, Synthesis and turnover of nitrate reductase induced by nitrate in cultured tobacco cells. J Biol Chem 246:1772-1779, 1971
99. Hageman, RH and D Flesher, Nitrate reductase activity in corn seedlings as affected by light and nitrate content of nutrient media. Plant Physiol 35:700-708, 1960

100. Travis, RL and JL Key, Correlation between polyribosome level and the ability to induce nitrate reductase activity in dark-grown corn seedlings. Plant Physiol 48:617-620, 1971

101. Beevers, L, LE Schrader, D Flesher and RH Hageman, The role of light and nitrate in the induction of nitrate reductase in radish cotyledons and maize seedlings. Plant Physiol 40:691-698, 1965

102. Jones, RW and RW Sheard, Phytochrome, nitrate movement and induction of nitrate reductase in etiolated pea terminal buds. Plant Physiol 55:954-959, 1975

103. Aslam, M, A Oaks and RC Huffaker, Effect of light and glucose on the induction of nitrate reductase and on the distribution of nitrate in etiolated barley leaves. Plant Physiol 58:588-591, 1976

104. Jolly, SO and NE Tolbert, NADH-nitrate reductase inhibitor from soybean leaves. Plant Physiol 62:197-203, 1978

105. Stewart, GR and D Rhodes, Control of enzyme levels in the regulation of nitrogen assimilation. In: Regulation of Enzyme Synthesis and Activity. Smith, H (ed.), London, Academic Press, 1-22, 1977

106. Funkhouser, EA and CS Ramadoss, Synthesis of nitrate reductase in *Chlorella*. II. Evidence for synthesis in ammonia-grown cells. Plant Physiol 65:944-948, 1980

107. Funkhouser, EA, T-C Shen and R. Ackermann, Synthesis of nitrate reductase in *Chlorella*. I. Evidence for an inactive protein precursor. Plant Physiol 65:939-943, 1980

108. Hipkin, CR and PJ Syrett, Post-transcriptional control of nitrate reductase formation in green algae. J Exp Bot 28:1270-1277, 1977

109. Meeker, GB, AC Purvis, CA Neyra and RH Hageman, Uptake and accumulation of nitrate as a major factor in the regulation of nitrate reductase activity in corn (*Zea mays* L.) leaves: Effects of high ambient CO_2 and malate. In: Mechanisms of Regulation of Plant Growth, Bieleski, RL, AR Ferguson and MM Cresswell (eds.), Bull 12, Wellington, The Royal Society of New Zealand, 49-58, 1974

110. Shaner, DL and JS Boyer, Nitrate reductase activity in maize (*Zea mays* L.) leaves. I. Regulation by nitrate flux. Plant Physiol 58:499-504, 1976

111. Ferrari, TE, OC Yoder and P Filner, Anaerobic nitrite production by plant cells and tissues: evidence for two nitrate pools. Plant Physiol 51:423-431, 1973

112. Canvin, DT and CA Atkins, Nitrate, nitrite and ammonia assimilation by leaves: effect of light, carbon dioxide and oxygen. Planta 116:207-224, 1974

113. Canvin, DT and KC Woo, The regulation of nitrate reduction in spinach leaves. Can J Bot 57:1155-1160, 1979

114. Sawhney, SK, MS Naik and DJD Nicholas, Regulation of NADH supply for nitrate reduction in green plants via photosynthesis and mitochondrial respiration. Biochem Biophys Res Commun 81:1209-1217, 1978

115. Sawhney, SK, MS Naik and DJD Nicholas, Regulation of nitrate reduction by light, ATP and mitochondrial respiration in wheat leaves. Nature 272:647-648, 1978

116. Woo, KC, M Jokinen and DT Canvin, Reduction of nitrate via a dicarboxylate shuttle in a reconstituted system of supernatant and mitochondria from spinach leaves. Plant Physiol 65:433-436, 1980

117. Aslam, M, RC Huffaker, DW Rains and KP Rao, Influence of light and ambient carbon dioxide concentration on nitrate assimilation by intact barley seedlings. Plant Physiol 63:1205-1209, 1979

118. Solomonson, LP and AM Spehar, Model for the regulation of nitrate assimilation. Nature 265:373-375, 1977

119. Tischner, R and A Hüttermann, Light mediated activation of nitrate reductase in synchronous _Chlorella_. Plant Physiol 62:284-286, 1978

120. Jones, RW and RW Sheard, Light factors in nitrogen assimilation. In: _Nitrogen Assimilation of Plants_. Hewitt, EJ and CV Cutting (eds.), London, Academic Press, 521-539, 1979

121. Klepper, L, D Flesher and RH Hageman, Generation of reduced nicotinamide adenine dinucleotide for nitrate reduction in green leaves. Plant Physiol 48:580-590, 1971

122. Nicholas, JC, JE Harper and RH Hageman, Nitrate reductase activity in soybeans (_Glycine max_ L.) Merr.). II. Energy limitations. Plant Physiol 58:736-739, 1976

123. Aparicio, PJ and JM Maldonado, Regulation of nitrate assimilation in photosynthetic organisms. In: _Nitrogen Assimilation of Plants_. Hewitt, EJ and CV Cutting (eds.), London, Academic Press, 207-215, 1979

124. Lorimer, GJ, H Gewitz, W Volker, LP Solomonson and B Vennesland, The presence of bound cyanide in the naturally inactivated form of nitrate reductase of _Chlorella vulgaris_. J Biol Chem 249:6074-6079, 1974

125. Losada, M, Metalloenzymes of the nitrate-reducing system. J Mol Catalysis 1:245-264, 1975/76

126. Syrett, PJ and JW Leftley, Nitrate and urea assimilation by algae. In: _Perspectives in Experimental Biology_. Sunderland, N (ed.), Oxford, Pergamon Press, 221-234, 1976

127. Ferguson, AR, The nitrogen metabolism of _Spirodela oligorrhiza_. II. Control of the enzymes of nitrate assimilation. Planta 88:353-363, 1969
128. Orebamjo, TO and GR Stewart, Ammonium repression of nitrate reductase formation in _Lemna minor_ L. Planta 122:27-36, 1975
129. Orebamjo, TO and GR Stewart, Ammonium inactivation of nitrate reductase in _Lemna minor_ L. Planta 122:37-44, 1975
130. Oaks, A, M Aslam and I Boesel, Ammonium and amino acids as regulators of nitrate reductase in corn roots. Plant Physiol 59:391-394, 1977
131. Radin, JW, Amino acid interactions in the regulation of nitrate reductase induction in cotton root tips. Plant Physiol 60:467-469, 1977
132. Filner, P, Regulation of nitrate reductase in cultured tobacco cells. Biochim Biophys Acta 118:299-310, 1966
133. Yamaya, T and K Ohira, Nitrate reductase inactivating factor from rice cells in suspension culture. Plant Cell Physiol 17:633-641, 1976
134. Yamaya, T and K Ohira, Nitrate reductase inactivating factor from rice seedlings. Plant Cell Physiol 19:211-220, 1978
135. Sherrard, JH, JA Kennedy and MJ Dalling, _In vitro_ stability of nitrate reductase from wheat leaves. II. Isolation of factors from crude extract which affect stability of highly purified nitrate reductase. Plant Physiol 64:439-444, 1979
136. Sherrard, JH, JA Kennedy and MJ Dalling, _In vitro_ stability of nitrate reductase from wheat leaves. III. Isolation and partial characterization of a nitrate reductase inactivating factor. Plant Physiol 64:640-645, 1979
137. Matsumoto, H, T Tanaka, T Matoh, K Hashizumi and E Takahashi, Inhibition of NADH-nitrate reductase activity in cucumber leaves due to NADH oxidation. Plant Cell Physiol 20:573-582, 1979
138. Yamaya, T, A Oaks, and IL Boesel, Characteristics of nitrate reductase-inactivating proteins obtained from corn roots and rice cell cultures. Plant Physiol 65:141-145, 1980
139. Yamaya, T, LP Solomonson and A Oaks, Action of corn and rice-inactivating proteins on a purified nitrate reductase from _Chlorella vulgaris_. Plant Physiol 65:146-150, 1980
140. Yamaya, T and K Ohira, Reversible inactivation of nitrate reductase by its inactivating factor from rice cells in suspension culture. Plant Cell Physiol 19:1085-1089, 1978
141. Schrader, LE, GL Ritenour, GL Eilrich and RH Hageman, Some characteristics of nitrate reductase from higher plants. Plant Physiol 43:930-940, 1968

142. Friedrich, JW and LE Schrader, Sulfur deprivation and nitrogen metabolism in maize seedlings. Plant Physiol 61:900-903, 1978
143. Onwueme, IC, HM Laude and RC Huffaker, Nitrate reductase activity in relation to heat stress in barley seedlings. Crop Sci 11:195-200, 1971
144. Alofe, CO, LE Schrader and RR Smith, Influence of high day and variable night temperatures on nitrate reductase activity of young corn (Zea mays L.) plants. Crop Sci 13:625-629, 1973
145. Huffaker, RC, T Radin, GE Kleinkopf and EL Cox, Effects of mild water stress on enzymes of nitrate assimilation and of the carboxylative phase of photosynthesis in barley. Crop Sci 10:471-474, 1970
146. Hageman, RH, Integration of nitrogen assimilation in relation to yield. In: Nitrogen Assimilation of Plants. Hewitt, EJ and CV Cutting (eds.), London, Academic Press, 591-611, 1979
147. Morilla, CA, JS Boyer and RH Hageman, Nitrate reductase activity and polyribosomal content of corn (Zea mays L.) having low leaf water potentials. Plant Physiol 51:817-824, 1973
148. Shaner, DL and JS Boyer, Nitrate reductase activity in maize (Zea mays L.) leaves. II. Regulation by nitrate flux at low leaf water potential. Plant Physiol 58:505-509, 1976
149. Deckard, EL, RJ Lambert and RH Hageman, Nitrate reductase activity in corn leaves as related to yields of grain and grain protein. Crop Sci 13:343-350, 1973
150. Eilrich, GL and RH Hageman, Nitrate reductase activity and its relationship to the accumulation of vegetative and grain nitrogen in wheat (Triticum aestivum L.). Crop Sci 13:59-66, 1973
151. Brunetti, N and RH Hageman, Comparison of in vivo and in vitro assays of nitrate reductase in wheat (Triticum aestivum L.) seedlings. Plant Physiol 58:583-587, 1976
152. Hucklesby, DP, CM Brown, SE Howell and RH Hageman, Late spring applications of nitrogen for efficient utilization and enhanced production of grain and grain protein in wheat. Agron J 63:274-276, 1971
153. Dalling, MJ, GM Halloran and JH Wilson, The relation between nitrate reductase activity and grain nitrogen productivity in wheat. Aust J Agric Res 26:1-10, 1975
154. Deckard, EL, KA Lucken, LR Joppar and JJ Hammond, Nitrate reductase activity, nitrogen distribution, grain yield, and grain protein of tall and semi-dwarf near-isogenic lines of Triticum aestivum and T. turgidum. Crop Sci 17:293-296, 1977
155. Johnson, CB, WJ Whittington and GC Blackwood, Nitrate reductase as a possible predictive test of crop yield. Nature 262:133-134, 1976

156. Blackwood, GC and R Hallam, Nitrate reductase activity in wheat (Triticum aestivum L.). II. The correlation with yield. New Phytol 82:417-425, 1979
157. Austin, RB, L Rossi and RD Blackwell, Relationships between nitrate reductase activity, plant weight and nitrogen content in seedlings of Triticum, Aegilops and Triticale. Ann Bot 42:429-438, 1978
158. Rao, KP, DW Rains, CO Qualset and RC Huffaker, Nitrogen nutrition and grain protein in two spring wheat genotypes differing in nitrate reductase activity. Crop Sci 17:283-286, 1977
159. Fakorede, MAB and JJ Mock, Nitrate-reductase activity and grain yield of maize cultivar hybrids, Crop Sci 18:680-682, 1978
160. Deckard, EL and RH Busch, Nitrate reductase assays as a prediction test for crosses and lines in spring wheat. Crop Sci 18:289-293, 1978
161. Pate, JS, Exchange of solutes between phloem and xylem and circulation in the whole plant. In: Encyclopedia of Plant Physiol 1. Transport in Plants. I. Phloem Transport, Zimmermann, MH and JA Milburn (eds.), Berlin, Springer-Verlag, 451-473, 1975
162. Pate, JS, Transport and partitioning of nitrogenous solutes. Annu Rev. Plant Physiol 31:313-340, 1980
163. Weissman, GS, Effect of ammonium and nitrate nutrition on protein level and exudate composition. Plant Physiol 39:947-952, 1964
164. Hill-Cottingham, DG and CP Lloyd-Jones, Translocation of nitrogenous compounds in plants. In: Nitrogen Assimilation of Plants, Hewitt, EJ and CV Cutting (eds.), London, Academic Press, 397-405, 1979
165. Martin, P, Pathways of uptake and translocation of nitrogen in kidney bean plants after uptake by the root. Z Pflanzenphysiol 64:206-222, 1971
166. Joy, KW and AJ Antcliff, Translocation of amino acids in sugar beet. Nature 211:210-211, 1966
167. Pate, JS, PJ Sharkey and CA Atkins, Nutrition of a developing legume fruit. Functional economy in terms of carbon, nitrogen, water. Plant Physiol 59:506-510, 1977
168. Housley, TL, DM Peterson, and LE Schrader, Long distance translocation of sucrose, serine, leucine, lysine, and CO_2 assimilates. I. Soybean. Plant Physiol 59:217-220, 1977
169. Huffaker, RC and LW Peterson, Protein turnover in plants and possible means of its regulation. Annu Rev Plant Physiol 25:363-392, 1974
170. Peterson, LW and RC Huffaker, Loss of ribulose-1,5 diphosphate carboxylase and increase in proteolytic activity during senescence of detached primary barley leaves. Plant Physiol 55:1009-1015, 1975

171. Wittenbach, VA, Ribulose bisphosphate carboxylase and proteolytic activity in wheat leaves from anthesis through senescence. Plant Physiol 65:884-887, 1979
172. Dalling, MJ, G Boland and JH Wilson, Relation between acid proteinase activity and redistribution of nitrogen during grain development in wheat. Aust J Plant Physiol 3:721-730, 1976
173. Feller, UK, TT Soong and RH Hageman, Leaf proteolytic activities and senescence during grain development of field-grown corn (Zea mays L.). Plant Physiol 59:290-294, 1977
174. Pate, JS, Nutrients and metabolites of fluids recovered from xylem and phloem: Significance in relation to long-distance transport in plants. In: Transport and Transfer Processes in Plants. Wardlaw, IF and JB Passioura (eds.), New York, Academic Press, 253-281, 1976
175. Thomas RJ, U Feller and KH Erismann. The effect of different inorganic nitrogen sources and plant age on the composition of bleeding sap of Phaseolus vulgaris (L.). New Phytol 82:657-669, 1979
176. McClure, PR and DW Israel, Transport of nitrogen in the xylem sap of soybean plants. Plant Physiol 64:411-416, 1979
177. Israel, DW and PR McClure, Nitrogen translocation in the xylem of soybeans. In: Proc World Soybean Res Conf II, Corbin, FT (ed.), Boulder, Westview Press, in press, 1980
178. MacRobbie, E, Phloem translocation. Facts and mechanisms. A comparative survey. Biol Rev. 46:429-481, 1971
179. Pate, JS, PJ Sharkey, and OAM Lewis, Xylem to phloem transfer of solutes in fruiting shoots of a legume, studied by a phloem bleeding technique. Planta 122:11-26, 1975
180. Hall, SM and DA Baker, The chemical composition of Ricinus phloem exudate. Planta 106:131-140, 1972
181. Nelson, CD, Translocation of organic compounds in plants. Can J Bot 40:757-770
182. Housley, TL, LE Schrader, M Miller and TL Setter, Partitioning of ^{14}C-photosynthate, and long distance translocation of amino acids in preflowering and flowering, nodulated and nonnodulated soybeans. Plant Physiol 64:94-98, 1979
183. Schrader, LE, TL Housley and JC Servaites, Amino acid loading and transport in phloem. In: Proc World Soybean Res Conf II, Corbin, FT (ed.), Boulder, Westview Press, in press, 1980
184. Garner, DCJ and AJ Peel, Metabolism and transport of ^{14}C-labeled glutamic and aspartic acids in the phloem of willow. Phytochemistry 10:2385-2387, 1971

185. Ziegler, H, Untersuchungen über die Reitung und Sekretion der Assimilate. Planta 47:447-500, 1956
186. van Die, J and PML Tammes, Phloem exudation from monocotyledonous axes. In: Encyclopedia of Plant Physiol 1. Transport in Plants. I. Phloem Transport. Zimmermann, MH and JA Milburn (eds.), Berlin, Springer-Verlag, 196-222, 1975
187. Duke, SH, LE Schrader, MG Miller and RL Niece, Low temperature effects on soybean (Glycine max [L.] Merr. cv. Wells) free amino acid pools during germination. Plant Physiol 62:642-647, 1978
188. Higgins, CF and JW Payne, Peptide transport by germinating barley embryos. Planta 134:205-206, 1977
189. Higgins, CF and JW Payne, Characterization of active dipeptide transport by germinating barley embryos: effects of pH and metabolic inhibitors. Planta 136:71-76
190. Clauss, H, DC Mortimer and PR Gorham, Time-course study of translocation of products of photosynthesis in soybean plants. Plant Physiol 39:269-273, 1964
191. Fisher, DB, Kinetics of ^{14}C translocation in soybean. I. Kinetics in the stem. Plant Physiol 45:107-113, 1970
192. Servaites, JC, LE Schrader and DM Jung, Energy-dependent loading of amino acids and sucrose into the phloem of soybean. Plant Physiol 64:546-550, 1979
193. Amar, L and L Reinhold, Loss of membrane transport ability in leaf cells and release of protein as a result of osmotic shock. Plant Physiol 51:620-625, 1973
194. Reinhold, L, RA Shtarkshall and D Ganot, Transport of amino acids in barley leaf tissue. J Exp Bot 223:809-817, 1970
195. Berlin, J and V Mutert, Evidence of distinct amino acid transport systems in cultured tobacco cells. Z Naturforsch 33:641-645, 1978
196. King, J and R Hirji, Amino acid transport systems of cultured soybean root cells. Can J Bot 53:2088-2091, 1975
197. Oxender, DL, Membrane transport. Annu Rev Biochem 41:777-810, 1972
198. Geiger, DR, Phloem loading. In: Encyclopedia of Plant Physiol 1. Transport in Plants. I. Phloem Transport. Zimmermann, MH and JA Milburn (eds.), Berlin, Springer-Verlag, 395-431, 1975
199. Giaquinta, RT, Evidence for phloem loading from the apoplast. Plant Physiol 57: 872-875, 1976
200. Giaquinta, R, Possible role of pH gradient and membrane ATPase in the loading of sucrose into the sieve tubes. Nature 267:369-370, 1977
201. van Bel, AJE, Lateral transport of amino acids and sugars during their flow through the xylem. Ph.D.

thesis, University of Utrecht, The Netherlands, 1978

202. Sinclair, TR and CT deWit, Photosynthate and nitrogen requirements for seed production by various crops. Science 189:565-567, 1975

203. Sinclair, TR and CT deWit, Analysis of the carbon and nitrogen limitations to soybean yield. Agron J 68:319-324, 1976

204. Penning deVries, FWT, Use of assimilates in higher plants. In: Photosynthesis and Productivity in Different Environments, Int Biol Prog 3, Cambridge, Cambridge Univ Press, 459-480, 1975

205. Garcia, RL and JJ Hanway, Foliar fertilization of soybeans during the seed-filling period. Agron J 68:653-657, 1976

206. Boote, KJ, RN Gallaher, WK Robertson, K Hinson and LC Hammond, Effect of foliar fertilization on photosynthesis, leaf nutrition, and yield of soybeans. Agron J 70:787-791, 1978

207. Sesay, A and R Shibles, Mineral depletion and leaf senescence in soya bean as influenced by foliar nutrient application during seed filling. Ann Bot 45:47-55, 1980

CHAPTER 4

CARBON AND NITROGEN PARTITIONING IN THE WHOLE PLANT - A THESIS BASED ON EMPIRICAL MODELING

John S. Pate and David B. Layzell
Department of Botany, The University of Western Australia, Nedlands, W.A. 6009

Introduction - The Basic Necessity for Differential Partitioning of C and N
Elements of Plant Functioning Likely to Foster Differential Partitioning of C and N
Experimentally Derived Data on C and N Balance of the Organs of a Plant and Implications to Plant Functioning
A Series of Simple, Empirical Representations of C and N Partitioning Based on the Data Obtained for White Lupin
- A Model Based on Mass Flow in Phloem
- A Model Based on Diffusion-based Flow (Non-mass Flow) in Phloem
- A Model Involving Mass Flow in Dual Phloem Streams of Differing Composition
- A Model Based on Mass Flow in Phloem Involving the Concept of a Superactive Upper Stratum of Leaves

A More Sophisticated Modeling Approach Including Information on C:N Ratios of Phloem Streams
Concluding Comments
Acknowledgements
References

J.D. Bewley (ed.), Nitrogen and Carbon Metabolism

INTRODUCTION - THE BASIC NECESSITY FOR DIFFERENTIAL PARTITIONING OF C AND N

Dry matter of plants contains a relatively constant proportion (38-46%) by weight of carbon and an amount of nitrogen (1-7%) which varies widely with plant species and with the type and age of plant from which it is derived. As shown in the examples given (Table 1), levels of N in dry matter tend to be higher in shoots than in roots, and, within shoots, to be higher in leaves than in stems and usually highest in shoot apices, young leaves and fruits. Values in excess of 4.5% N may signify the presence in a tissue of concentrated forms of protein, such as the chloroplastic proteins of leaves and the protein bodies of seeds and certain other reserve organs. In mature seeds of legumes such as soybean (*Glycine max*), the cotyledons may contain 40% or more of their dry matter as protein, with N comprising up to 7% of the dry weight. Roots rarely store protein to such a high level even when forming storage tap roots or tubers, the value of 11% protein in fresh weight recorded for root tubers of winged bean (*Psophocarpus tetragonolobus*) (29) being probably close to the maximum and well above the average attained by root crops.

As well as partitioning newly assimilated C and N, the plant engages continuously in a series of transfers of previously accumulated C and N between its constituent parts and tissues. Throughout vegetative development of the annual crop plant this mobilization occurs principally from old to young age groups of leaves and from old to young parts of stem and root, but, once reproduction has commenced mobilization takes place from all of these vegetative parts to fruits or to other harvestable structures such as tubers and bulbs. As a general rule N is retrieved much more effectively from senescent

Table 1. Percentage of nitrogen in dry matter of below-ground and shoot parts of various crop plants

Species (age in days at sampling)			% N in Dry Matter		Reference Number
			Below-Ground Parts[1]	Shoot or Shoot Parts[2]	
White lupin	*(Lupinus albus)*	(51-58d)	2.7(NR)	3.7(S); 4.7(SA)	(36)
Narrow-leaved lupin	*(Lupinus angustifolius)*	(70d)	2.0(NR)	5.0(L); 2.1(SP)	(11)
Cowpea	*(Vigna unguiculata)*	(80d)	1.4(NR)	2.4(S); 3.3(F)	(17)
Cowpea	*(Vigna unguiculata)*	(49d)	2.4(NR)	3.0(S); 4.5(L); 1.5(SP)	(9)
Soybean	*(Glycine max)*	(43d)	2.4(NR)	3.2(S); 4.5(L)	(64)
Soybean	*(Glycine max)*	(40d) (symbiotic)	2.4(NR)	3.3(S)	(77)
Soybean	*(Glycine max)*	(40d) (nitrate-grown)			
Broad bean	*(Vicia faba)*	(60d)	1.5(NR)	3.8(S); 5.0(L)	(5)
Peanut	*(Arachis hypogaea)*	(100d)	1.5(NR)	3.4(L); 4.2(F)	(76)
Cotton	*(Gossypium hirsutum)*	(56d)	2.6(R)	3.0(S); 3.6(L)	(64)
Tobacco	*(Nicotiana tabacum)*	(31d)	1.6(R)	2.8(S); 3.5(L)	(65)
Barley	*(Hordeum vulgare)*	(42d)	1.2(R)	3.4(L)	(12)

[1]R, root; NR, nodulated root;

[2]S, whole shoot; L, leaves; SP, stem and petioles; F, flowers and/or fruits; SA, shoot apex

plant tissues than is C, mainly because the N is bound largely into protein, which is fully degradable at senescence, whilst the C is associated largely with cell walls whose fabric is essentially left intact when an organ dies or is shed from the parent plant. The C of starch, sugars and higher oligosaccharides such as fructosans is potentially mobilizable at organ senescence, but in most non-harvested parts of monocarpic crop plants these forms of carbohydrate comprise a relatively small reservoir of carbon compared to that bound in cellulose and lignin.

A principal outcome of the differential retrieval of C and N from the vegetative parts of a grain crop is that the C requirement for fruit formation and filling derives largely from the photosynthesis which accompanies reproduction, whereas the N requirement for seed formation can be met to a large extent from N assimilated before flowering and mobilized from root, stem and leaf as fruiting proceeds (e.g. 10, 54, 57). Accordingly, the proportion of total plant N which is harvested as seed at the end of the growth cycle of a productive grain crop is likely to be as high as 65-75%, whereas the proportion of the total C of a plant present in its seed is unlikely to exceed 40-45%. As great or greater differential mobilization of N and C is observed in developing fruits. The seeds of legumes, for example, commonly contain 80% of the N of the mature fruit but less than half of the carbon (59). The consumption of carbon by plant tissue involves release of CO_2 in respiration as well as the incorporation of carbon into new dry matter. It has long been known that rates of dark respiration tend to be highest in youngest, actively growing parts of the plant, and decline markedly with age of tissue (e.g. see 19, 30, 70). In the experimental plant used in this chapter, the grain legume (*Lupinus albus* L. cv. Ultra), the apical regions of the shoot, principally young lateral apices and the terminal inflorescence, exhibit a dark respiration rate (CO_2 evolution per unit dry matter) two to three times that of mature stem tissue. Also, as shown in Table 2, it appears typical of certain legumes to have significantly greater dark respiratory losses per unit dry matter of root (nodulated or non-nodulated) than per unit dry matter of shoot. Extending the comparison to dark respiration rates of root and young apical regions of the shoot, the shoot apex turns out to be generally the faster consumer of C in respiration, although this is not always the case (see

Table 2. Relative rates of respiration of above- and below-ground parts of various plants species

Species (age in days at sampling)				Ratio[1] Shoot Respiration / Root Respiration	Reference Number
White lupin	*(Lupinus albus)*	(51-58d)		0.6(S); 0.6(L); 0.4(SP); 1.4(SA)	This study
White lupin	*(Lupinus albus)*	(75d)		0.6(S)	(55)
Cowpea	*(Vigna unguiculata)*	(60d)		0.6(S)	(17)
Cowpea	*(Vigna unguiculata)*	(25d)		0.7(S)	(68)
White clover	*(Trifolium repens)*	(45d)		0.6(S)	(68)
Soybean	*(Glycine max)*	(40d)	(nitrate-grown)	0.5(S)	(67)
Soybean	*(Glycine max)*	(40d)	(symbiotic)	0.3(S)	(67)
Tobacco	*(Nicotiana tabacum)*			10.9(L); 3.5(SP)	(66)
Sunflower	*(Helianthus annuus)*			1.7(L); 0.9(SP)	(22)
Apple	*(Pyrus malus)*			2.4(L); 2.3(SP)	(61)
Potato	*(Solanum tuberosum)*			7.0(SA)	Pate, unpublished
Cucumber	*(Cucumis sativus)*			2.8(SA)	Pate, unpublished
Sow thistle	*(Sonchus arvensis)*			0.6(SA)	Pate, unpublished
Castor bean	*(Ricinus cummunis)*			4.1(SA)	Pate, unpublished

[1]Calculated as the specific activity (dry weight basis) of respiration of shoot or shoot parts ÷ the specific activity (dry weight basis) of respiration of below ground parts (roots + nodules (if present)). S, entire shoots; L, leaves; SP, stems and/or petioles; SA, stem apex.

exceptions in Table 2). However, if the green tissue of the shoot apex photosynthesizes during the photoperiod, CO_2 losses will be ameliorated if not abolished during this time. Indeed photosynthetic reassimilation of respired CO_2 is likely to have considerable impact on the carbon economy of a shoot under growing conditions in which there is a large differential in day to night temperature. For example in our data for *Lupinus albus*, for which a ratio of shoot apex dark respiration rate to root dark respiration rate of 1.4 was recorded (Table 2), the actual loss of C as CO_2 recorded for the shoot apex per 24 hour period per unit dry matter turned out to be less than half that of the nodulated root. This was because the shoot apex was maintained at or close to CO_2 compensation point during the photoperiod whilst the root was respiring at a much higher than average rate because of a 12°C difference between day and night temperature.

Before leaving the topic of respiratory loss of C, it is worth mentioning that organs fully heterotrophic for C and N are likely through their respiratory losses to produce dry matter with a substantially lower C:N ratio than in the translocate they derive from other parts of the plant. Certain roots exhibiting high overall respiratory activity have been shown to exhibit a cyanide-resistant component in their respiration, characterized by being non-productive of ATP and therefore apparently 'wasteful' in conventional biochemical terms (32, 33). By maintaining high rates of carbon loss in this manner, the root would continue to command high sink activity and thereby gain more N as phloem translocate than would be attracted with a lesser respiratory consumption of carbon. Advantage might accrue from this alternative chain form of respiration especially in situations in which availability of assimilated N from the shoot was determining the rate of root growth.

Having established that the parts of a plant differ widely in C:N ratio of their dry matter, and bearing in mind that they differ greatly in their extent of autotrophy for C and N, one reaches the obvious conclusion that plant organs are likely to vary very widely in their respective requirements for C and N. In view of this one would expect that mechanisms are likely to have evolved in plants for dispensing C and N to a large extent independently of one another through the transport channels of the plant body. The purpose of this chapter is to determine whether such mechanisms exist, and, if so, to suggest how they operate and interact during plant growth and development.

ELEMENTS OF PLANT FUNCTIONING LIKELY TO FOSTER DIFFERENTIAL PARTITIONING OF C AND N

Although the below-ground parts of plants may participate in assimilation of inorganic forms of N such as NH_4^+, NO_3^- and N_2 it is rare for the resulting organic solutes of N to accumulate at sites of assimilation. Instead, assimilatory products are quickly released to the xylem and thence immediately available to the shoot in the transpiration stream (26, 46). Indeed, so effective is upward export that portions of the root distal to sites of N assimilation (e.g. root tissue below nodules in the case of legumes fixing N, and apical tips of roots in plants feeding on NO_3) must depend largely on N which cycles through the shoot and back to these distal regions as phloem translocate (e.g. see $^{15}N_2$ and $^{15}NO_3$ feeding experiments of Oghoghorie and Pate (46)).

Xylem fluid usually contains large amounts of N-rich solutes such as ureides and amides as its dominant organic constituents and has a C:N weight ratio usually less than 3:1 (48, 50), so that its intake by shoot organs would provide a most effective means of meeting the relatively large requirement of the shoot for N relative to C. Transpiring leaves are not the only organs to benefit directly from the nitrogenous solutes of xylem, since stem tissue can engage in lateral uptake of N from the ascending xylem stream (18, 58) and fruits and shoot apex exhibit measurable, though admittedly low, rates of transpiration, and are therefore likely to acquire some N as xylem fluid (56).

A second process which militates in favour of enrichment of shoots with N is the capacity of the root to load the xylem stream with forms of N other than those arising from current assimilatory activity. These non-assimilatory components of xylem N are viewed as arising from catabolic processes within the root, from the release of stored soluble forms of N from root tissues, or from the return to the shoot of N donated to the root as phloem translocate. The evidence for these types of activities has been recently discussed (50); suffice to say that in the few quantitive investigations which have been made it appears that in the normal, healthy vegetative plant this non-assimilatory component of xylem N rarely accounts for more than 10-15% of the N currently being exported from the root. In certain plants the non-assimilatory component of xylem appears to consist of a different spectrum of N compounds to that typical of primary assimilation of N. In tomato (*Lycopersicon esculentum*), for example, valine and isoleucine/leucine are regarded as catabolic products released to xylem whilst glutamine is the principal product of N assimilation (7); in pea (*Pisum sativum*), aspartic acid, valine and leucines are principally components not formed in quantity from current N assimilation (47, 75); and in nodulated *Lupinus albus*, in which root xylem exudates have been collected above and below the zone of nodules on a root system, aspartic acid, valine, and leucine/isoleucine are, for the most part, the non-assimilatory components of xylem N, whilst the bulk of the asparagine + glutamine reaching the shoot can be traced to the recent fixation of N by root nodules (D.B. Layzell and J.S. Pate, unpublished).

A third element of differential partitioning of C and N relates to the phloem translocate exported from leaves. The literature on crop plants reveals many instances in which the photosynthate produced by lower, older strata of leaves is directed to roots and lower parts of the shoot, whilst the photosynthate of upper younger leaves serves the terminal growing parts of the shoot system (6, 31, 74). While this arrangement achieves obvious economies in terms of minimizing distances over which solutes are transported, and possible expediencies in terms of the integration of source and sink activities, it is likely to work against rather than in favour of the enrichment of shoot sinks with N. This is because older, lower leaves tend to export translocate richer in N relative to C than do younger upper leaves, presumably since old leaves are likely to be less active in photosynthetic fixation of C, and more prone to lose N as they approach senescence than their younger counterparts higher up the shoot. Direct evidence of this comes from studies in which phloem exudates have been collected from petioles or veins of leaves of different age on a plant; assuming of course that the exudates so obtained offer a reliable sample of the translocate currently being generated within the leaf. Earlier data provided for lupin (56) and that in this paper, bear evidence of quite substantial down gradients in C:N ratio of phloem sap from young to old leaves of this species. A similar phenomenon is observed in *Ricinus communis* (J.S. Pate and D.B. Layzell, unpublished), and the equivalent effect in woody deciduous species is found at the approach of leaf senescence in autumn when phloem exudates become significantly more enriched with N relative to C than at any time during the previous summer (78).

In summary, then, we suggest that it may be general amongst plants for growing regions of the shoot to display a much lower requirement for C relative to N than do roots. This difference arises from the higher % N in dry matter of shoot apex than root, the generally high respiratory losses of C by roots, and the ability of young shoot tissues, if green and adequately illuminated, to compensate for day-time respiratory loss by photosynthesis. The transpiration of the shoot emerges as the dominant activity whereby assimilated and non-assimilated forms of N are released from the root, and the phloem translocation stream emanating from photosynthesizing leaves as the major agency of delivery of C and cycled or mobilized N to the various centres of growth or storage within the plant. Although the roots of many species may have the potential to meet part or all of their N requirement by direct incorporation of products of N assimilation, and shoot organs have the capacity to be more or less self-supporting for C through photosynthetic activity, the translocatory processes of the plant more than compensate for these nutritional advantages and foster a partitioning of solutes, which, on the whole, transfers N preferentially to the shoot and C preferentially to the below-ground parts of the plant. To determine how this is accomplished requires detailed information on the C and N economy of a species as the following sections will testify.

EXPERIMENTALLY DERIVED DATA ON C AND N BALANCE OF THE ORGANS OF A PLANT AND IMPLICATIONS TO PLANT FUNCTIONING

Few accounts are found in the literature on the uptake and utilization of C and N by plant organs, and, to the best of our knowledge, no quantitatively explicit descriptions are available on organ-to-organ interchange of C and N in relation to assimilation and growth of a whole plant other than a legume. At cellular level the situation regarding the integration of C and N metabolism is more hopeful, with particularly useful studies now appearing on the siting of the key reactions of N assimilation (44), on the interaction of NO_3 reduction and CO_2 fixation in photosynthesizing tissues (2), and on N turnover in relation to photorespiration of plants (4, 27). These types of information need to be carefully integrated for the tissues of individual species before the metabolism of C and N at organ and whole plant level can be satisfactorily appreciated. Otherwise the modeling of C and N partitioning in plants becomes a somewhat pointless exercise in which inadequate knowledge combined with false assumptions can lead to a most misleading picture of how plants function.

The approach in our laboratory has been to construct detailed profiles of production, usage and transport of C and N within a whole plant, based on studies of the economies of C, N and H_2O in specific plant parts over clearly defined intervals of plant growth. For purposes of this review we will select a set of data relating to the 10-day period after the commencement of flowering of nodulated plants of *Lupinus albus*, as compiled in a recent study of the species (36). The functional units of the plant whose activities are followed during the study interval are indicated in Figure 1. The use of nodulated plants, relying solely on atmospheric nitrogen, may appear somewhat unusual, but has the advantage of examining a situation in which two mutually dependent assimilatory functions, photosynthesis and N fixation, are occurring in fully autotrophic context, unaffected by any external supply of inorganic N. Note that four separate strata (age groups) of leaflets (L_1-L_4) (Figure 1) and of stem and petioles (SP_1-SP_4) are distinguished, leaflets being treated separately from other shoot parts in view of their intense activity in photosynthesis and transpiration. The principal centres of extension growth at this stage of plant development are the nodulated root (NR) and the apical growing regions of the shoot (A), the latter comprising the terminal inflorescence and the young lateral shoots developing at the top three or four nodes of the main stem. Leaflets of L_4 though not all fully expanded at the commencement of the study period and still accumulating C and N, were shown to be past the stage of importing assimilates through the phloem and therefore regarded as being totally dependent on xylem for their intake of N. The information obtained included,

1. changes in C and N content of plant parts as determined from dry weight and analysis of C and N in dry matter,
2. estimates of water consumption of shoot parts from measurements of losses or gains of tissue water, estimates of water used metabolically, gravimetric measurements of transpirational loss of water from shoots of whole intact plants, and assessments of the proportional contributions of individual shoot parts to this

transpiration by short-term gravimetric measurement of water loss of freshly harvested shoot parts.

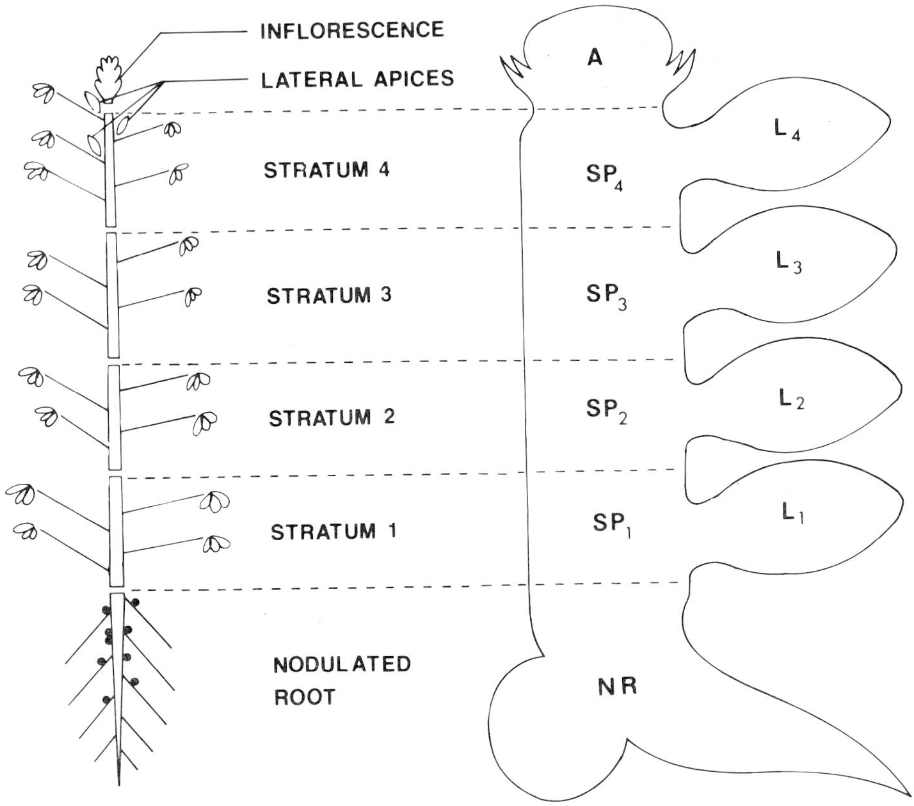

Figure 1: Diagrammatic representations of a plant of Lupinus albus *for the period 51 to 58 days after sowing to show stage of development (left) and groupings of plant organs (right) used when constructing the various models of C and N partitioning. A, apex (including inflorescence and lateral apices); SP_1 to SP_4, stem + petioles of shoot strata 1 to 4; L_1 to L_4, leaflets of shoot strata 1 to 4; NR, nodulated root.*

3. determinations of respiratory losses of C from nodulated roots by collection and assay of the respired CO_2 from enclosed root systems of intact plants,
4. measurements of night respiration of whole shoots and assessment of the proportional contributions of shoot parts to this respiration

by short-term measurements of respiratory loss of freshly detached shoot parts, by infra red gas analysis,

5. indirect estimates of the net photosynthetic gain of C by the whole shoot by summation of (a) the increment of C by the whole plant in dry matter, (b) the day and night respiration loss of C by the nodulated root, and (c) the night respiratory loss of C by the shoot system,
6. assessments of the proportional contributions of the different strata of shoot segments to net daytime fixation of C by measurement of CO_2 exchange of attached shoot parts by infra red gas analysis, and by a short-term $^{14}CO_2$ feeding technique,
7. examination of the source:sink relationships for photosynthate by ^{14}C feeding of leaflets of the four leaf strata and studying the subsequent distribution of the resulting ^{14}C assimilates by phloem sap analysis or radioassay of parts of the fed plants, and
8. assay of solute composition and C:N weight ratio of solutes of xylem bleeding sap from roots and of phloem exudates collected from shallow incisions in stem or petioles at a series of strategically-located positions on the shoot.

The techniques used when collecting the above information are described in detail elsewhere (36, 56). The modeling exercises which follow this section will make use of all of these classes of data, with the exception of information on phloem sap composition which is not utilized until a final, more sophisticated version of the modeling procedure is presented.

The amounts of C, N and H_2O consumed by the various parts of our lupin plants are represented pictorially in Figure 2, in which sizes (areas) of plant organs are drawn in proportion to the relative amounts of a specific commodity which they utilized during the 10-day study period. The large differences in overall 'shape' of these 'cartoon-like' representations indicate substantial differences in patterns of utilization of the three substances by plant organs. Thus, if expressed as percentages of the total amount of C, N or H_2O acquired by the plant during the study period, stem + petioles consumed 22% of the C, but only 12% of the N and 5% of the H_2O, whilst leaflets utilized 27% of the C, 40% of the N and 93% of the H_2O. Of the strata of shoot segments L_3, L_4 and SP_4 were large consumers of N; L_2 and L_3 greatest transpirers of H_2O; L_3, L_4 and SP_3 the three high consumers of C.

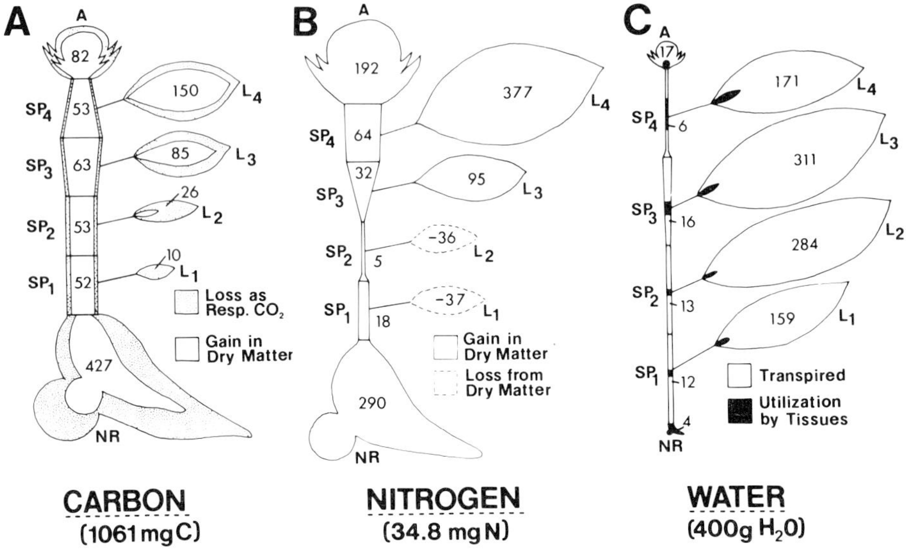

Figure 2: Consumption of C, N and H_2O by plant parts of 51 to 58 day Lupinus albus. The relative areas occupied by, and the numbers associated with, plant organs or groups of organs are given proportional to a consumption by the whole plant of 1000 units by weight of C, N or H_2O. Absolute amounts consumed by the entire plant over the study period are indicated below each figure. Plant parts designated as in Figure 1.

Respiratory losses are represented in Figure 2 as day + night loss of C (as CO_2) from the nodulated root, whilst respiration of shoot parts refers only to night time respiration, as all shoot organs were maintained at or above CO_2 compensation point during daytime. Losses of C as CO_2 from each plant part are represented as stippled areas of the profiles for carbon utilization (Figure 2A), with corresponding unstippled areas denoting consumption of C in net synthesis of dry matter. Note the large proportional loss of C as CO_2 by the nodulated root relative to the amount of C incorporated into root dry matter, and the substantial loss of respired CO_2 from lower leaflets (L_1 and L_2) which were showing net losses of N or of C and N from their dry matter. At this stage of development the main stem still comprised a major sink for C, although all but its top segment (SP_4) had ceased to elongate.

The data highlighted the very substantial difference between shoot apex (A) and nodulated root (NR) in relative requirements for C and N, the shoot apical parts acquiring 19% of the total N assimilated but only 8% of the C of net photosynthate, the nodulated root 29% of the N and as much as 43% of the C.

The above differences in organ requirement are better displayed in terms of weight ratio of C consumed to N consumed, as shown in Figure 3. Shoot apex (C:N intake of 13:1) and upper leaflets (12:1) had highest requirement for N relative to C, and old stem tissue engaging in secondary thickening highest relative requirement for C (137C:1N). Roots exhibited a four-fold greater requirement of C relative to N (54C:1N) than did apical regions of the shoot (12-13C:1N), a finding which reinforces the general statements on C and N requirements of shoot and root made at the end of the previous section.

Figure 3 includes data for photosynthetic inputs of C by the four strata of leaflets, these parts being the only organs making significant net gains of C from the atmosphere during the photoperiod. As was to be expected, leaflets of L_2 and L_3 were more active photosynthetically than the youngest (L_4) and oldest (L_1) strata of leaflets. Taken together the leaflets assimilated 30 units by weight of C for every unit by weight of N fixed by the root nodules, an overall input ratio for the whole plant of 30:1, against which the requirements of individual organs for C and N can be matched.

The C:N ratio of xylem sap was 2:1 (see Figure 3) stressing the likely importance of the xylem in increasing the intake by shoot parts of N relative to C. Also included is information on the pattern of flow of photosynthate from leaves, as indicated by ^{14}C feeding. By this stage of development assimilate distribution was already highly stratified with L_4 leaflets committed exclusively to upward translocation to the shoot apical parts and adjacent stem tissue, L_3 leaflets feeding all parts of the plant, and L_1 and L_2 leaflets supplying photosynthate entirely in a downward direction to lower stem and root system. With the lower half of the stem functioning as a relatively small sink for N, any N mobilized from lower leaves was likely to pass mainly to the root system.

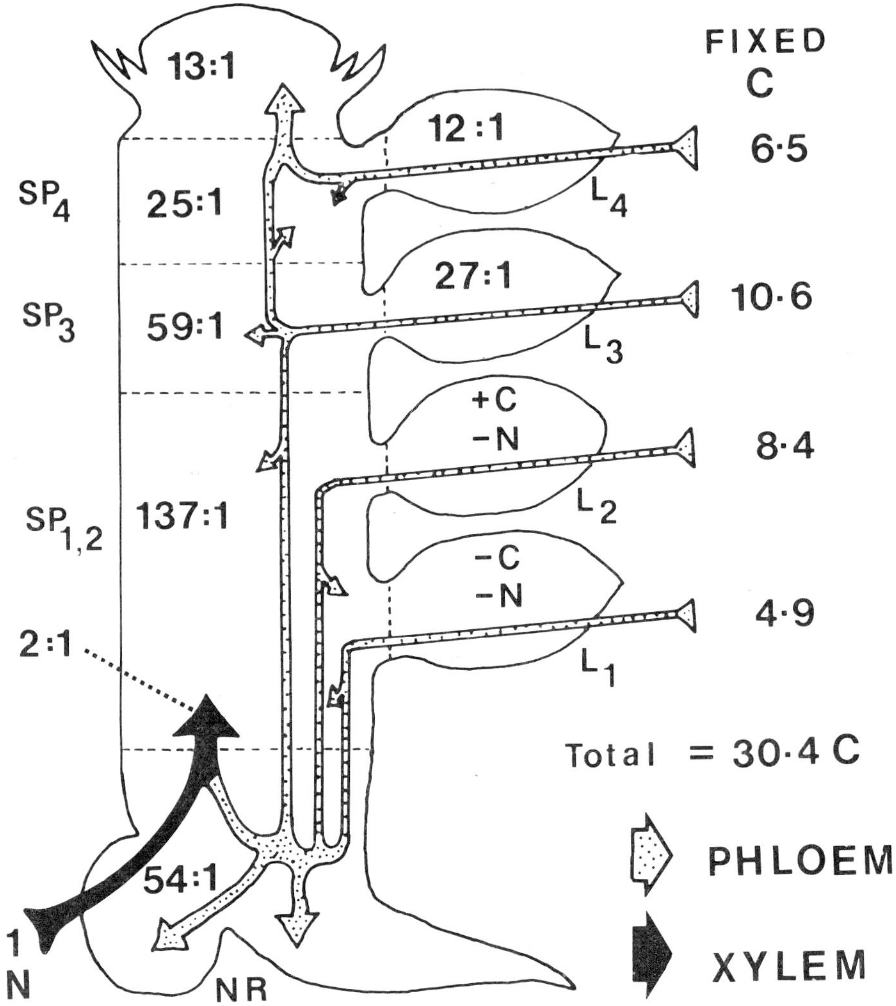

Figure 3: Diagrammatic representation of 51 to 58 day Lupinus albus plant showing the C:N weight ratio of consumption (including C lost in respiration) of each plant organ or stratum of organs. The relative amounts of C (as net photosynthate) fixed by each leaflet strata are indicated relative to the assimilation of 1 unit of N through nitrogen fixation by root nodules. The C:N requirement of the nodulated root includes the amount of C leaving the root with N in the xylem (C:N ratio = 2:1. Plant parts designated as in Figure 1.

A SERIES OF SIMPLE, EMPIRICAL REPRESENTATIONS OF C AND N PARTITIONING BASED ON THE DATA OBTAINED FOR WHITE LUPIN

Using the information presented in Figures 2 and 3, a series of quantitatively-based models may be constructed to represent a series of possible solutions to the problem of how differential transport of C and N might be achieved within the plant body. Each model is based on certain fundamental assumptions on how the transport processes of the plant might operate and on how the various parts of the plant co-operate in exchanges of C, N and H_2O.

All of the models to be considered assume that long-distance transport occurs exclusively through xylem and phloem, thereby making the concepts expressed inapplicable to species possessing alternative vehicles of transport such as by means of lactifers or resin or mucilage ducts. All models assume that xylem transport occurs in a strictly upward direction from roots to centres of transpiration in the shoot, with C and N moving in xylem in the proportions suggested from analysis of xylem sap. The carbon incorporated into the nitrogenous compounds exported from roots in the xylem is pictured as coming from the C of net photosynthate translocated to the nodulated root during the study period. All models attempt, wherever possible, to incorporate the measured values on C, N and H_2O utilization by plant parts, and, in all but one case, conform to the data on transpirational loss and photosynthetic activity by the four strata of leaflets.
All modeling exercises considered at this stage assume that the stem abstracts C and N from xylem and phloem to meet its own requirements for growth and respiration, but does not have the capacity to engage in transfer of C and N between different strands and conducting tissues of its vasculature. Thus, at this stage in our thesis, the stem is regarded as having a purely passive role in the partitioning process.

All models subscribe to 'mass flow' of C, N and H_2O in xylem, and, in all but one representation, the mass flow mechanism of translocation is selected as the means whereby exchanges of C, N and H_2O are effected between source and sink organs through the phloem. And, finally, all but one of the models assume that leaflets import exclusively through xylem and export exclusively through phloem, and that the streams of phloem translocate which each group of leaflets generates is distributed within the plant body essentially as indicated in the ^{14}C feeding studies (see Figure 3). Each model will now be considered in turn.

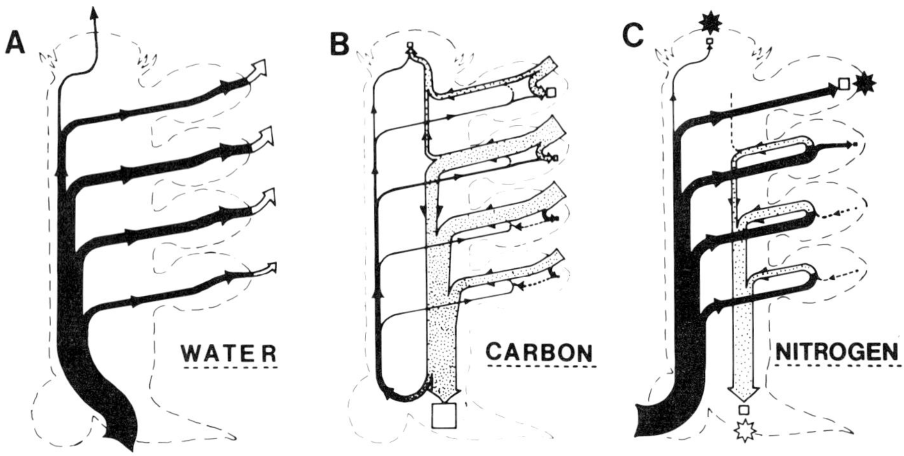

Figure 4: Model of uptake, flow and utilization of C, N and H_2O based on mass flow in phloem. The models fit the recorded data for C and H_2O consumption, but the derived model for N flow undersupplies the shoot apex and upper stratum of leaflets (black stars) whilst grossly oversupplying the root (open star). Black lines - xylem flow; stippled lines - phloem flow. Lines are drawn of thickness relative to the amounts of C, N or H_2O moving through different pathways. The boxes representing utilization of C, N or H_2O by plant parts are drawn relative to amounts of a commodity consumed by each part (see text for further details).

A Model Based on Mass Flow in Phloem (Figure 4).

This simple model is built to conform to all data for C and H_2O utilization of plant parts. Water (Figure 4A) is pictured as passing to shoot apex and leaflets in proportion to their measured transpirational losses, whilst inputs of C from the leaflet strata are shown to be distributed by the phloem in the directions indicated by ^{14}C feeding and in amounts which meet precisely the recorded consumption of C by plant parts. The attendant distribution of N is constrained to follow that for phloem flow of C, and, taken together with the complementary flow of N in xylem based on mass flow with water in the transpiration stream, there results the model for N flow depicted in Figure 4C. This flow profile for N is clearly inadequate since the nodulated root receives through phloem from the shoot more than twice its recorded increment of N, whilst the shoot apex and upper stratum of leaflets receive through the xylem only 7% and 46% respectively of the N they were recorded as having consumed during the study period. Furthermore, in this representation (Figure 4) the translocate from upper leaves consists entirely of carbon, since the xylem input of N by upper leaflets is grossly insufficient to match the requirements of these leaflets for N, let alone to generate an exportable surplus of N for nourishment of the adjacent shoot apex.

The above inadequacies might be somewhat ameliorated were one to include the possibility of a return flow to the shoot *via* the xylem of some of the excess N received by the root as phloem translocate. According to the N model (Figure 4C) a portion (18%) of any such cycled N would be drawn off into the upper strata of leaflets whilst the remainder (82%) would return to the root after cycling through the lower strata of leaves. One would then presume that this N would be available for further cycles of passage between root and shoot and thereby for further progressive abstraction by the top region of the shoot. Extrapolating to the situation in which the cycling process were to furnish a sufficient draw-off of N to satisfy the recorded intake of N by shoot apex and upper leaflets, it can be estimated that over 90% of the N leaving the root at any time in the xylem would have to be in the process of cycling, and that the equivalent of some 80% of the current content of total N in the plant body would then have to be in the transport channels of the plant. These are clearly most unreasonable assumptions for white lupin, since less than 15% of the plants' total N consists of free amino compounds (52) and labeling studies and collections of xylem exudate above and below the zone of nodules on a root have suggested that, at most only 10-15% of the N currently in xylem is in the process of cycling between root and shoot system (52). Faced with these facts, this form of model becomes unacceptable.

A Model Based on Diffusion-based Flow (Non-mass Flow) in Phloem (Figure 5)

Many authors have pointed to the inherent inability of a translocation process based on mass flow to distribute solutes released from a source in accordance with the very different demands of sink organs for specific solutes or sets of solutes. Herein lies the attraction of invoking a mechanism of phloem transport in which some form of facilitated diffusion is the driving force for solute transfer, and in which specific solutes can move more or less independently of one another along their own solute potential gradients (e.g. see (21, 71)).

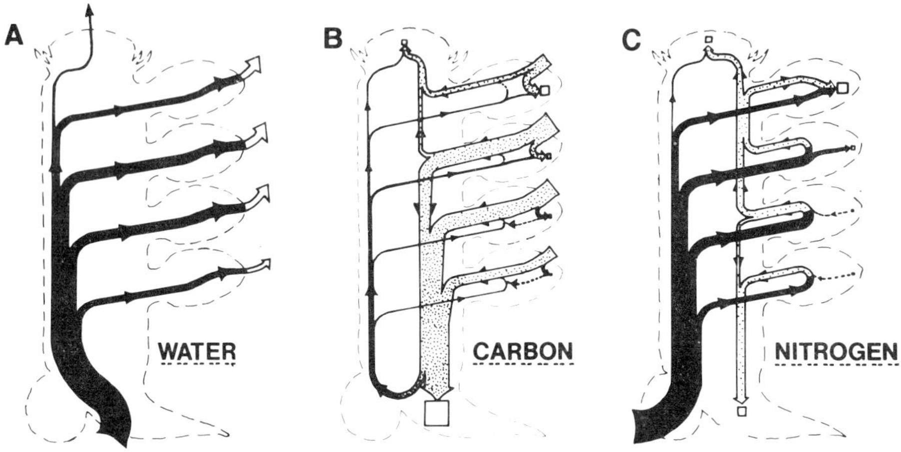

Figure 5: Model of partitioning of C, N and H_2O based on diffusion-based flow in phloem. The model fits the recorded consumption of C, N and H_2O in plant parts, but has certain unsatisfactory features which are detailed in the text. Black lines - xylem flow; stippled lines - phloem flow. Lines of flow drawn in thickness proportional to amounts of traffic of C, N, or H_2O along the different transport pathways.

Applying a diffusion-based concept of this kind to the present data, one arrives at what appears on first sight to be a fully adequate partitioning system (see Figure 5). This model incorporates identical patterns of flow of C and H_2O, as in the previous case, but permits, through the diffusion analogue, the generation of a flow pattern for N in phloem totally independent of that of C and therefore able to achieve the desired objective of fulfilling the recorded increments of N in all plant parts.

There are, however, several unacceptable features regarding the operation of phloem transport. Firstly, the model concludes that bidirectional transport occurs in the phloem of the petioles of the uppermost stratum of leaves, an outgoing stream of phloem-borne C passing out of these leaves as an opposing stream of phloem-borne N enters to satisfy the N increment of the leaflets. Secondly, when the translocate from L_2 leaflets reaches the stem it is pictured as being split into two very different streams, one set of solutes very rich in C relative to N diffusing downwards to the root, another, exclusively of N, passing upwards to satisfy the high requirement of the upper regions of the shoot for N. Thirdly, if one compares the suggested rates of transfer of C and N to root and upper parts of the shoot through the phloem, translocation to roots is found to involve an up to 25 times higher transfer of C relative to N than is in upward phloem transfer to the shoot apex. Were this so, there would have to be a difference of similar magnitude in the rates of loading of N & C onto phloem in the respective nurse leaves to root and shoot apex, and unrealistically steep concentration gradients for specific solutes would have to be generated in order to maintain such an unequal partitioning to sink organs. The mathematical bases for the above criticisms have been set out and discussed by several authors (e.g. (1, 60)), rendering the diffusion-based model much less attractive than one might suppose on purely conceptual grounds.

A Model Involving Mass Flow in Dual Phloem Streams of Differing Composition (Figure 6)

This model differs from the earlier one based on mass flow in phloem (Figure 4) by assuming that leaves are capable of producing two distinct streams of phloem translocate, differing widely in C:N ratio and destined to supply different sink regions with assimilates. The model (Figure 6) satisfies the measured consumption of C, N and H_2O of plant parts by imagining that nitrogen-rich streams of translocate from the lower three-leaf strata feed the shoot apex and upper stratum of leaves, and that phloem streams considerably richer in C relative to N move downwards from all age groups of leaves to supply the nodulated root.

The feature of this model which one might wish especially to challenge is the supposition that massive phloem import of N takes place into upper leaves, just as these same leaves are engaging in equally active phloem export of C-rich translocate to roots. In all published accounts of changes in direction and intensity of phloem translocation during the development of leaves, the phase when export and import through phloem is observed is of short duration and completed well before the leaf is fully expanded (19, 20, 24, 73). In this model, however, bidirectional translocation is assumed to continue through to the adult life of the leaf.

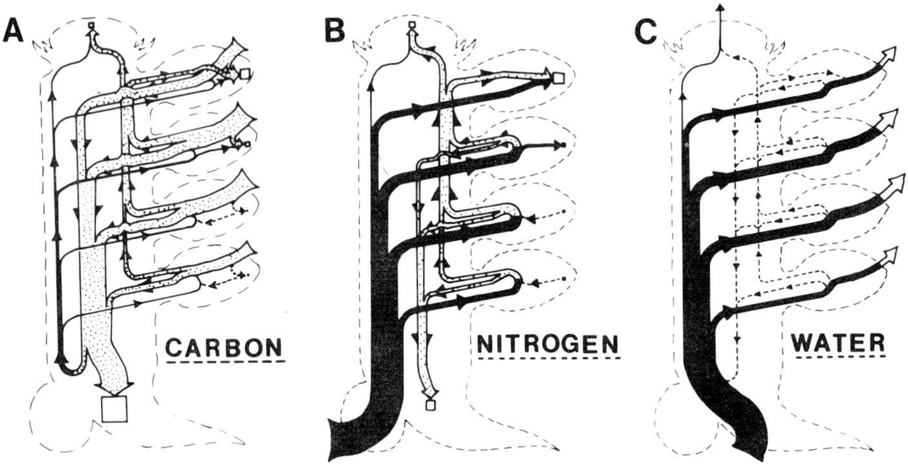

Figure 6: Model of C, N and H_2O partitioning based on the concept of dual phloem streams of differing composition supplying different sink regions with the required amounts of C and N. The model satisfies the recorded consumptions of C, N and H_2O by plant parts but note bidirectional transport in phloem in petioles of the top stratum of leaves and the wide differences in C:N ratio of phloem streams. Black lines - xylem flow; stippled lines - phloem flow (see text for further details).

A dual phloem stream model fails to conform to the evidence of distribution of ^{14}C assimilates in *Lupinus albus* (see Figure 3), so is unlikely to apply to this species or to any other with similarly organized nutritional commitments amongst the leaves of its shoot. The model is likely to be relevant, however, to species such as tomato in which ^{14}C-feeding studies have shown that both acropetal and basipetal flow of assimilates can occur from a fed leaf, and that this is attributable to the presence of internal as well as external phloem (3, 28). Apical parts of the shoot of this species appear to receive some translocate from lower leaves *via* the external phloem, the roots their translocate largely *via* the internal phloem. In fruiting plants an individual truss of fruits can obtain assimilate from virtually any leaf (28). These are the very sorts of translocatory arrangements which one would expect from the partitioning system suggested in Figure 6.

Anatomical studies (see 43) have shown that some 27 families of Dicotyledons exhibit 'intraxylary' or internal phloem, suggesting that the system of assimilate transport suggested in Figure 6 might be quite widespread in plants, including members of economically-important families such as *Asteraceae*, *Convolvulaceae*, *Myrtaceae*, *Euphorbiaceae* and *Solanaceae*. Whatever the nature of their transport exchanges, plants possessing internal and external phloem appear to have an inherent plasticity in source:sink relationships not likely to be present in a plant with only a conventional network of external phloem, and detailed knowledge of their translocatory arrangements is awaited with interest.

A Model Based on Mass Flow in Phloem and Involving the Concept of a Superactive Upper Stratum of Leaves (Figure 7)

The basic problem with each of the models examined so far is one of distributing sufficient N to the shoot apex and youngest stratum of leaves. If simultaneous import and export through phloem of the upper leaves is not regarded as an acceptable proposition, the logical alternative is to propose that xylem is responsible for delivering the necessary N. One way in which this might be accomplished is for upper leaflets to transpire at an inordinately fast rate and in so doing to attract sufficient N through the xylem to provide not only their own needs but also sufficient additional N to enable these leaflets to feed the adjacent apical region of the shoot through the phloem. This apical region is unlikely to attract much xylem fluid since it has a low surface area to volume ratio, has its young organs enveloped in developing leaves and hairs, and is likely to lack stomata.

The model which results from the concept of a superactive stratum of upper leaflets (Figure 7), suggests that almost two-thirds of the transpiration of the shoot would have to reside in the uppermost stratum of leaflets in order to implement the N requirements of the top of the shoot by intake of xylem fluid. This represents a proportional activity in transpiration by L_4 leaflets some three-and-a-half times that observed experimentally (cf Figures 2C and 7B).

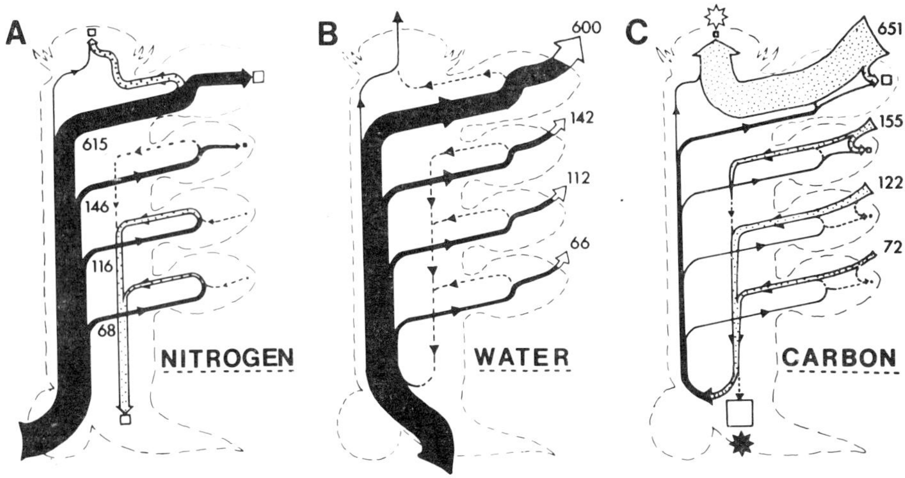

Figure 7: Model of C, N and H_2O partitioning based on the concept of a superactive stratum of leaves transpiring sufficiently fast to attract enough N to satisfy the recorded increments of N in upper leaflets and apical region of shoot. Mass flow is assumed to occur in xylem and phloem. The derived model for C flow assumes photosynthetic inputs of leaf strata to be commensurate with transpirational losses. Note that the resulting model for C partitioning grossly oversupplies (open star) the apical region of the shoot and grossly undersupplies (black star) the root with C. Numbers in the figure refer to proportioning of 1000 units by weight of C, N or H_2O. Black lines - xylem flow; stippled lines - phloem flow (see text for further details).

Our studies of white lupin have shown that a very close correlation exists between transpirational loss of water and photosynthetic fixation of CO_2 over a wide range of environmental conditions and age groups of expanded leaves (C.A. Atkins and J.S. Pate, unpublished), so it is reasonable to suppose that the photosynthetic input of the upper leaflets would increase commensurate with the increase in transpirational activity proposed in the model shown in Figure 7. There would then result a profile for C partitioning which fails to match the observed C increments of plant parts, the shoot apex ending up grossly oversupplied with C and the root grossly undersupplied with C (see Figure 7C). This misfit, and the dubious assumption that upper leaves might achieve rates of exchange of H_2O and CO_2 with the surrounding atmosphere several times greater than has been observed experimentally for *L. albus*, forces one to abandon the model. Nevertheless, some of its features might well be relevant to the situation in over-dense crop canopies in which uppermost leaves would be the only ones sufficiently well-illuminated for near maximum rates of CO_2 fixation and water loss. Whilst plant growth rate might be reduced considerably under such circumstances, the resulting partitioning of C and N might well approach closely to that suggested in Figure 7. Indeed, it has been shown that roots of closely-spaced plants suffer particularly from shortage of assimilates (34, 39), as the model would have us believe, and our own studies of densely-sown potted legumes have shown unusually high proportions of N to be concentrated in the top region of the shoot, again as might be envisaged from the model of Figure 7.

A MORE SOPHISTICATED MODELING APPROACH INCLUDING INFORMATION ON C:N RATIOS OF PHLOEM STREAMS (Figures 8-10)

In most crop species phloem fails to bleed when vascular tissue is cut, and the collection of phloem contents using aphids is not a feasible proposition. However, in our experimental species, *Lupinus albus*, in certain other herbaceous genera (e.g. *Pisum, Brassica, Matthiola, Antirrhinum, Ricinus*), in a range of woody species (79), and in certain arborescent Monocotyledons (8), phloem exudates can be obtained from cut phloem, and in certain of these plants (e.g. *Ricinus* and *Lupinus*) phloem bleeds at virtually any location on the shoot, thereby enabling phloem streams to be tapped at a range of points in the plant's translocatory system (45, 51). With information on solute composition and, particularly, on C:N ratios of phloem sap from these various locations, a completely new dimension is added to the modeling of the exchanges of C and N between specific organs and the rest of the plant through xylem and phloem. This section sets out to demonstrate what can be achieved in this direction.

Figure 8 illustrates typical data for C:N weight ratios of phloem exudates collected at a range of sites on vegetative shoots of *Lupinus albus* plants identical to those employed for compiling the information on C, N and H_2O economy used in the models described earlier in the chapter. Figure 8 shows that the upper stratum of leaves exported translocate richest in C relative in N, while lower leaves produced translocate least rich in C. Translocate collected from phloem of

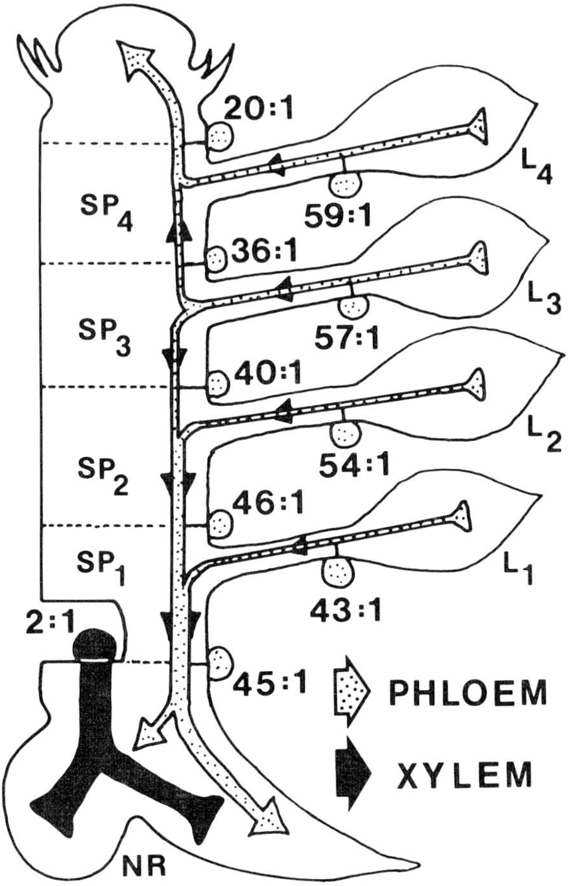

Figure 8: Diagrammatic representation of 51 to 58 day Lupinus albus showing the C:N weight ratio of transport fluids collected at specific sites throughout the plant body. Plant parts designated as in Figure 1. A model drawn using this data is shown in Figure 9 (see text for further details).

stem tissue exhibited the reverse of this trend with much lower C:N values for exudates collected close to the shoot apex than in exudates obtained from the stem base. Superimposing onto these findings the translocatory pathways for assimilates indicated from ^{14}C feeding of leaves (see Figure 8), it is concluded that the phloem translocate passing to roots contains approximately the same balance of C to N as is present in the assimilate streams emanating from the donor leaves, whilst translocate passing into apical regions of the shoot reaches the top of the stem carrying a considerably greater amount of N relative to C than was present in the original phloem streams produced by the relevant strata of upper leaves.

Comparisons of solute concentrations in upper petiole phloem sap with those in upper stem phloem sap have been made on several occasions in *Lupinus albus* (51, 56), and these consistently suggest that the observed lowering of the C:N ratio of phloem translocate in its upward passage through the shoot is achieved by addition to the phloem of the upper stem of certain key nitrogenous solutes, particularly the amides asparagine and glutamine. Sucrose concentrations are somewhat lower in stem top phloem exudates than in petiole phloem sap of the donor leaves, and since this difference in concentration is approximately equal, on a molar basis, to the concomitant increase in level of amino compounds, it is concluded that the upper segments of the shoot are capable of unloading sucrose from the passing phloem stream and loading an equivalent amount of amide (51). This exchange lowers significantly the C:N ratio of the phloem stream and thus provides the shoot apex with a translocate whose N content is much closer to the growth requirement of apical tissues than were these tissues to rely solely on translocate derived from the upper leaves. At the same time, the removal of sucrose from the ascending phloem stream provides the expanding stem and petiole segments of the upper region of the shoot with their primary source of C.

Preliminary data on *Ricinus communis* (J.S. Pate and D.B. Layzell, unpublished) indicate that this species might possess a very similar partitioning system in its stem to that of *Lupinus*. As shown in an earlier study (15) of *Ricinus*, glutamine is the principal component of xylem and phloem, and our results indicate that this solute is also responsible for enriching with N the phloem sap of the upper stem.

Returning to the original data for white lupin, it can now be appreciated how information on phloem sap composition makes the modeling exercise much more definitive and productive. Information on C:N ratios of petiole phloem sap, when combined with data on photosynthetic inputs of C by leaflets and changes in C and N content of leaflets, allows one to estimate precisely how much N is cycling through each stratum of leaves. By combining data on C:N ratio of stem base phloem sap with data on the consumption of C by roots, the amount of N supplied to the root in the phloem can be computed. Comparing this amount with the actual increment of N in the root, one can then assess to what extent N might be cycling through the root system. And, by comparing the C and N requirement of apical regions of the shoot with C:N ratios of xylem sap and stem top phloem sap, a determination can be made of the proportional intake by the shoot apex of C and N through xylem and phloem.

The modeling procedure becomes even more tightly constrained if, as shown in Figure 8, information is incorporated on the C:N ratios in phloem and xylem sap collected at intermediate points up the stem. Any change in C:N ratio in transport streams can then be equated with the measured requirement of the relevant petioles and stem segments for C and N, and a check thus made of whether the observed changes in C:N ratios of phloem streams are quantitatively compatible with the observed uptake of C and N by the shoot segments through which these streams of assimilates are passing.

A model incorporating this information on phloem sap composition, and satisfying fully the experimental observations for uptake and utilization of C, N and H_2O by plant parts is illustrated in Figure 9.

The reader is referred to two recent papers (36, 56) for discussion of the mathematical procedures used in modeling, and the possible sources of error and variability when constructing empirical representations of this nature. Suffice to say that the model conforms to mass flow in xylem, utilizes all data on C, N and H_2O utilization by plant parts and on net photosynthesis and transpiration of shoot parts, and sets out to achieve the objective of meeting the requirements of all plant parts for C and N by exchanges of solutes through xylem and phloem in the proportions of C and N suggested from analysis of the nearest relevant sampling point for xylem or phloem sap.

The resulting carbon budget (Figure 9A) depicts the C attached to amino compounds of xylem as coming from phloem translocate from the shoot, and, as in earlier models (Figures 4-7) simplifies the situation relating to carbon by combining respiratory losses of C and increments of C in dry matter as single 'box'-type representations depicting the total amount of C consumed by each plant organ. The water budget (Figure 9C) depicts water moving with C and N by mass flow in phloem and xylem, and water flow in phloem taking place as indicated from concentrations of solutes in phloem sap. The nitrogen budget (Figure 9B) suggests that N arriving in phloem to the root surplus to the root's measured N increment is cycled back to the root in the xylem, and that the equivalent of half of the N increment of the root nodules comes from direct incorporation of fixed N (35).

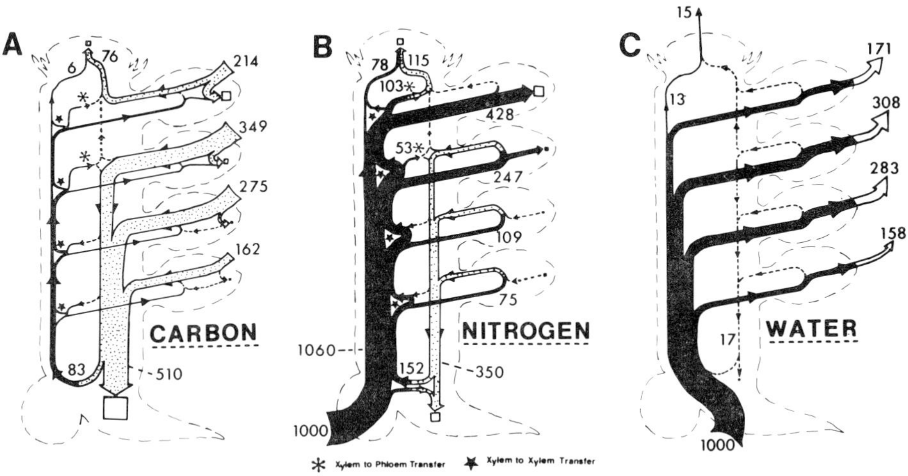

Figure 9: Sophisticated model of partitioning of C, N and H_2O in Lupinus albus based on information on C:N weight ratios of xylem and phloem (Figure 8) and meeting the measured consumptions of C, N and H_2O by plant parts (Figure 2). Mass flow is assumed to occur in xylem and phloem. Note suggested participation of stem in xylem to phloem and xylem to xylem transfers. Numbers refer to proportioning of 1000 units by weight of C, N or H_2O. Black lines - xylem flow; stippled lines - phloem flow (see text for further details).

The feature which sets this model apart from the others discussed earlier is that a successful fit of the data forces one to conclude that stem tissue has the capacity to engage in a series of exchanges of C and N between its different transport channels.

Firstly, the model indicates that a quite massive transfer of N occurs from xylem to phloem in the top half of the shoot and that this is the process largely responsible for decreasing the C:N ratio of phloem translocate as it passes to the shoot apex. It can be seen from Figure 9B, that 'xylem to phloem' transfer provides the shoot apex with a much larger fraction of its N intake than might be possible from direct uptake of xylem fluid, or from a release of N from upper leaves in the phloem.

Secondly, the model envisages that a xylem to xylem transfer system of some magnitude operates within the stem, this being indicated in Figure 9 by the looped pathways marked by black stars in the xylem pathways for N and C. This conclusion is reached by comparing the fluxes of H_2O (Figure 9C) and N (Figure 9B) at specific levels up the shoot and in the petioles of successive strata of leaves. The mainstream of xylem fluid in the body of the stem is then predicted to be (36) significantly more concentrated in N than that passing through petioles to leaves at the same level of the shoot, and there is a general tendency for xylem streams to be less concentrated in N in the lower than in upper parts of the shoot (36). This progressive concentration of N in xylem from base to top of shoot can be detected experimentally by vacuum extraction of xylem sap from freshly-excised lower, mid and upper segments of actively transpiring shoots, the tracheal sap from the upper stem being up to 2-3 times higher in concentration of nitrogenous solutes than in sap collected from the base of the stem (D.B. Layzell and J.S. Pate, unpublished).

This process of 'xylem to xylem' transfer has been suggested (36) to consist of a removal of N solutes from the xylem streams of vascular traces committed to supplying leaves, and the subsequent feedback of this N, or an equivalent amount of N, into xylem of adjacent stem traces moving into the next internode and destined to supply higher regions of the shoot. The model suggests that this retrieval of N from leaf traces occurs at all levels of the shoot, and in proportions indicated by the relative thickness of the looped pathways depicted for each stratum of shoot. It has been suggested that xylem parenchyma transfer cells lining the departing leaf traces are probable sites of uptake of N from xylem and that the vascular traces closing the 'gaps' caused by the departure of these traces are the most likely location for subsequent transfer of N to xylem streams moving further up the stem (4, 14, 36, 42).

We may now re-examine the central problem of how mature parts of the shoot effect a differential partitioning of C and N to shoot apex and nodulated root system. This is best displayed by using the data of Figure 9 to construct a picture of the various source agencies for transfer of C and N to shoot apex and root, indicating the relative contributions which each of these agencies makes to these two sink regions of the plant. This approach (Figure 10) highlights the role of xylem to phloem transfer in nourishment of the shoot apex with N, and the complementary role of upper leaf translocate in providing the C consumed by the apex.

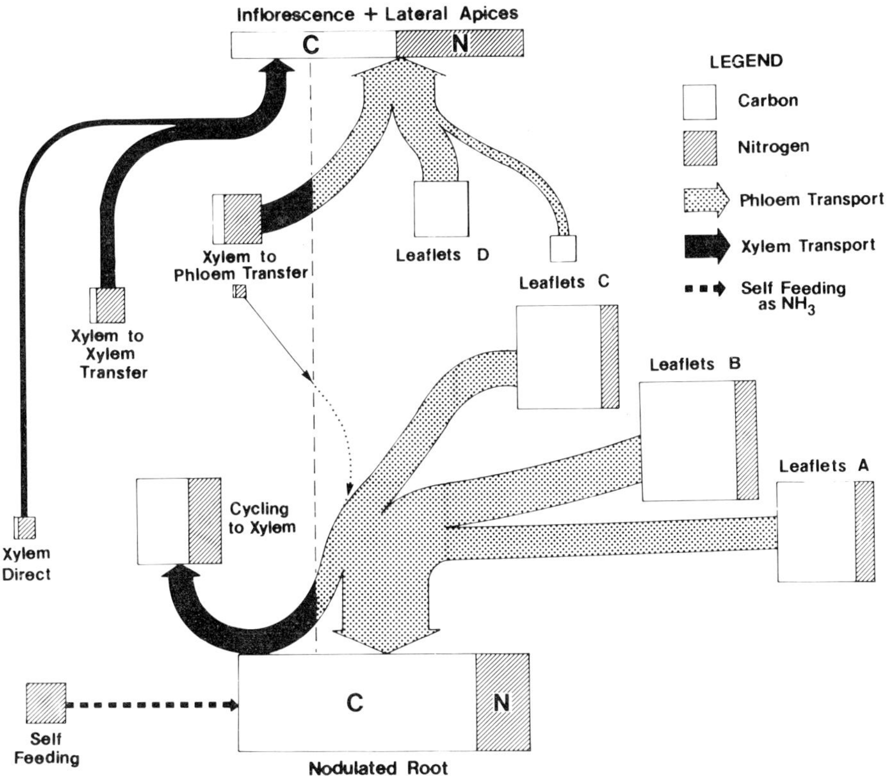

Figure 10: Diagrammatic representation of the source agencies of the mature shoot which supply C and N to the young apical region of the shoot (inflorescence + lateral apices) and the nodulated root. Relative amounts of C and N supplied from each source to the two sink regions are indicated using flow lines of varying thickness, and rectangles (sinks) or squares (sources) of varying area. An area scale is used in which 10 units by weight of C are equivalent in area to 1 unit by weight of N. Model derived from Figure 9 (see text and (36) for further details).

The model (Figure 10) distinguishes two 'hypothetical' components of xylem transport to the shoot apex, one, labelled 'xylem direct', pictured as coming from uptake of xylem fluid in an amount predicted were there to be no progressive enrichment of N in xylem from base to top of the shoot, the other, labelled 'xylem to xylem' transfer, resulting from withdrawal of N from leaf traces and feedback to stem xylem as suggested from the model of N flow in Figure 9.

The much greater intake of C relative to N by nodulated root than shoot apex is then seen to result from the relatively high C:N ratio of phloem translocate from lower leaves, yet this translocate and the N fixed directly into the dry matter of root nodules (marked 'self-feeding' in Figure 16) furnishes more than sufficient N for below-ground parts. This surplus of N is envisaged as cycling back to join the recently fixed N exported from root nodules in the xylem, thereby providing a further source of N for the shoot system.

CONCLUDING COMMENTS

As far as we are aware, the foregoing account for *Lupinus albus* L. is the first instance in which a detailed, experimentally-based analysis has been made of the transport exchanges of C and N through xylem and phloem in relation to the growth and functioning of a whole plant. Although it is not yet possible to know how widely the partitioning processes displayed by white lupin apply to other crop plants, the substantial usage by this species of stem tissue for exchanges of solutes between transport channels, and the derivation, thereby, of transport streams enriched with N for supply to apical parts of the shoot, may well be a strategy widely selected by plants. In this context, the nodal regions of stems are pictured as active centres for vascular exchanges, and believing that the presence of transfer cells might belie such activity (13, 14), the apparent ubiquity of these specialized cells in nodal tissue of Monocotyledons and Dicotyledons lends support to the suggestion that the strategic use of stem tissue for partitioning solutes might be widespread amongst flowering plants.

So far the partitioning process has been considered in a herbaceous Dicotyledon with relatively straightforward anatomical connections between plant parts, and a pattern of assimilate distribution in which an upper stratum of photosynthetic organs serves the growing termini of the shoot system, while lower and mid strata of leaves serve the roots. As mentioned earlier in relation to plants with external and internal phloem, not all Dicotyledons conform to the above pattern, and in Monocotyledon crop plants, such as cereals, extreme complexities may exist in nodal anatomy and in the organization of assimilate flow between leaves, shoots and tillers (10, 16, 40, 62).

As well as exhibiting wide variations in vascular anatomy, crop plants differ widely in N metabolism. Some reduce nitrate almost exclusively in their shoots, particularly in photosynthesizing leaves (e.g. *Gossypium, Capsicum)*, others *(Raphanus, Pisum, Lupinus)* are particularly active in assimilating NO_3 in their roots, and many others exhibit reduction in both root and shoot, with proportional activities of these organs in reduction depending on age and nutrition (48, 50, 63, 69). Some crop plants utilize ureides for transport of N in xylem from roots (e.g. many tropical legumes) (25, 53), others, such as cereals and temperate legumes, utilize amides (50). Some crop species store significant amounts of soluble N as free nitrate (see 41, 69), others tend to accumulate soluble N in reduced form, often as the ureides or amides mentioned above (52, 72). The impact of such variations on C and N partitioning is still poorly understood.

In a plant subsisting solely on NO_3-N and capable of reducing NO_3 only or almost exclusively in its leaves, all or a very large proportion of the reduced N necessary for protein synthesis in other organs has to be supplied as translocate from leaves. Composition of phloem sap of such species would be of interest, as would information on their source:sink relationships for translocate. With leaf rather than root as primary source of amino N, the requirement for massive exchanges of N between transport channels within the stem might well be reduced, whilst, were the root unable to reduce NO_3 at sufficiently fast rate to meet its own growth requirement for N, phloem transport of reduced N from the shoot would exercise an obligatory role in root nutrition and growth. However, this is unlikely to be the case in plants with active reductase capacity in below-ground parts, since relatively little reductase activity appears to be necessary to meet the moderate requirement of the root for N (see data for cotton (63)).

A second feature of influence on C and N partitioning relates to the form of N moving in xylem and phloem of a species. Judging from studies on legumes, tomato and apple, the anionic forms of N in xylem (e.g. allantoic acid, NO_3^-, aspartic acid and glutamic acid) are likely to be prevented by co-ion exclusion from exchange reactions with cell walls and protoplasts of the xylem and therefore will tend to be delivered directly to leaves in the transpiration stream. On the other hand, basic forms of N (e.g. arginine, NH_4^+) and compounds such as allantoin, glutamine and asparagine, which are uncharged or only minimally ionized at the pH of xylem sap, are seemingly more readily available for xylem to xylem or xylem to phloem transfer in stems (50). Since most species which have been investigated carry a range of amino compounds and often some free NO_3 in xylem sap, heterogeneity in behaviour of these compounds in the transport exchanges of the plant may be viewed as of primary importance in regulating the initial partitioning of N from the root (42). The nutritional implications of this remain to be evaluated.

The metabolism of xylem-delivered N in the shoot and the release of soluble forms of N for phloem loading represent a complex series of interrelated events of ultimate significance in determining the spectrum of nitrogenous solutes which accompanies sugar in the phloem. Compartmentation phenomena, varying rates of metabolism within leaf mesophyll, differential release to sites of phloem loading, and discrimination between different forms of N at the actual loading sites are all mechanisms which might determine what forms of N enter phloem, but we are still woefully ignorant of these processes and of their programming in cells and tissues. Present evidence suggests that free nitrate is not present in quantity in phloem sap even when present in large amount in xylem of a species (38, 50), but only a few species have been investigated in this respect. In plants assimilating NO_3 in their roots or subsisting on fixed N from root nodules, the phloem stream tends to consist of a wide mixture of amino compounds, some apparently resulting from direct xylem to phloem transfer in the minor veins of the leaf, others arising during leaf photosynthesis. In synthesizing its photosynthetic complement of amino compound the leaf breaks down the N-rich solutes such as asparagine, glutamine and allantoin which it receives *via* the xylem (37, 48, 49). Clearly, much more attention should be devoted to study of phloem loading of C and N in leaves, particularly in relation to the effects of leaf age and nutritional condition on the ratio of sugar to amino compounds in the leaf's translocate. Recent studies on soybean

(23, 69) have set the pattern for future studies of this nature and should be extended to other species. Once more is understood of what regulates the loading of C relative to N in phloem, the more will one be able to appreciate how partitioning is controlled during plant development.

ACKNOWLEDGEMENTS

It is a pleasure to record the technical assistance in this study of K. Hamel, E. Rasins, D. Waldie and P. Sanford. We are also greatly indebted to Dr. C. Atkins, Professor D.T. Canvin, and Dr. D. McNeil for their participation experimentally (see reference list) and through discussion in the modeling concepts developed in this review.

REFERENCES

1. Aikman, DP and DC Wildon, Phloem transport - surface flow hypotheses. J Exp Bot 29:387-394, 1978

2. Atkins, CA and DT Canvin, Nitrate, nitrite and ammonia assimilation by leaves: effects of inhibitors. Planta 123:41-51, 1975

3. Bonnemain, MJL, Sur le transport diurne des products d'assimilation lors des la foraison chez la tomate. Compt Rend 260: 2054-2057, 1965

4. Brown, RH, A difference in nitrogen use efficiency in C_3 and C_4 plants and its implications in adaptation and evolution. Crop Science 18:93-98, 1978

5. Cooper, DR, DG Hill-Cottingham and CP Lloyd-Jones, Absorption and re-distribution of nitrogen during growth and development of field beans, *Vicia faba*. Physiol Plant 38(4):313-318, 1976

6. Crafts, AS and CE Crisp, Phloem Transport in Plants. WH Freeman & Co., San Francisco, 1971

7. Die, J van, Synthesis and translocation of glutamine and related substances in the tomato plant. Proc Kon Ned Akad Wetensch, Series C 64:375-381, 1961

8. Die, J van, and PML Tammes, Phloem exudation from monocotyledonous axes. In: Encyclopedia of Plant Physiology I. Transport in Plants, I. Phloem Transport (Zimmermann, MR and JA Milburn eds). Springer, Berlin, 1975

9. Eaglesham, ARJ, FR Minchin, RJ Summerfield, PJ Dart, PA Huxley and JM Day, Nitrogen nutrition of cowpea (*Vigna unguiculata*) III. Distribution of nitrogen within effectively nodulated plants. Exp Agric 13:369-380, 1977

10. Evans, LT, IF Wardlaw and RA Fischer, Wheat, In: "Crop Physiology, some case histories" (Evans, LT ed) Cambridge University Press pp 101-149, 1975

11. Farrington, P, EAN Greenwood, ZV Titmanis, MJ Trinick and DW Smith, Fixation, accumulation and distribution of N in a crop of *Lupinus angustifolius* cv Unicrop. Aust J Agric Res 28:237-48, 1977

12. Goodall, DW and FG Gregory, Chemical composition of plants as an index of their nutritional status. Technical communication No 17. Imp Bureau Hort and Plantation Crops (Publ), Aberystwyth, Wales pp 49-53, 1947

13. Gunning, BES, Transfer cells and their roles in transport of solutes in plants. Sci Progr 64:539-68, 1977

14. Gunning, BES and JS Pate, Transfer cells, In: "Dynamic Aspects of Plant Ultrastructure" (Robards, AW ed). McGraw Hill, London pp 441-480, 1974

15. Hall, SM and DA Baker, The chemical composition of *Ricinus* phloem exudate. Planta 106:131-140, 1972

16. Hartt, CE and HP Kortschak, Tracing sugar in the cane plant. Proc. 11th Congress of the ISSCT Mauritius, 1962. Published as Paper No 97 in Journal Series of the Experimental Station, Hawaiian Sugar Planters' Ass, Honolulu, Hawaii, 1963

17. Herridge, DF and JS Pate, Utilization of net photosynthesis for nitrogen fixation and protein production in an annual legume. Plant Physiol 60:759-64, 1977

18. Hill-Cottingham, DG and CP Lloyd-Jones, Relative mobility of some organic nitrogenous compounds in the xylem of apple shoots. Nature 220:389-390, 1968

19. Ho, LC and AF Shaw, Carbon economy and translocation of ^{14}C in leaflets of the seventh leaf of tomato during leaf expansion. Ann Bot 41:833-848, 1977

20. Ho, LC and AF Shaw, Net accumulation of minerals and water and the carbon budget of an expanding leaf of tomato. Ann Bot 43:45-54, 1979

21. Honert, TH van den, On the mechanism of transport of organic materials in plants. Proc Kon Ned Akad Wetensch 35:1104-1111 cited in Sutcliffe and Collins (1975), 1932

22. Horie, T, Simulation of sunflower growth. I. Formulation and parameterization of dry matter production, leaf photosynthesis, respiration and partitioning of photosynthates. Bull Nat Inst Agric Sci, Series A 24:45-69, 1977

23. Housley, TL, LE Schrader, M Miller and TL Setter, Partitioning of ^{14}C-photosynthate, and long distance translocation of amino acids in preflowering and flowering, nodulated and non-nodulated soybeans. Plant Physiol 64:94-98, 1979

24. Isebrands, JG, RE Dickson and PR Larson, Translocation and incorporation of ^{14}C into the petiole from different regions within developing cottonwood leaves. Planta 128:185-193, 1976

25. Israel, DW and PR McClure, Nitrogen translocation in the xylem of soybeans, In: Proceedings World Soybean Research Conference II Rayleigh, North Carolina, USA, 1979

26. Ivanko, S, Metabolic pathways of nitrogen assimilation in plant tissues when ^{15}N is used as a tracer, In: "Nitrogen-15 in Soil Plant Studies", International Atomic Energy Agency, Vienna, 1971

27. Keys, AJ, IF Bird, MJ Cornelius, PJ Lea, RM Wallsgrove, BJ Miflin, Photorespiratory nitrogen cycle. Nature 275:741-743, 1978

28. Khan, AA and GR Sagar, Distribution of ^{14}C-labelled products of photosynthesis during the commercial life of the tomato crop. Ann Bot 30:727-743, 1966

29. Khan, TN and GE Eagleton, The winged bean, *Psophocarpus tetragonolobus*, In: Advances in Legume Science (Summerfield, RJ and AH Bunting eds), Royal Botanic Gardens, Kew, England, 1980

30. Kidd, F, C West and GE Briggs, A qualitative analysis of the growth of *Helianthus annuus*, Part 1, The respiration of the plant and of its parts throughout the life cycle. Proc Roy Soc B 92:368-84, 1921

31. Kursanov, AL, Metabolism and the transport of organic substances in the phloem, In: "Advances in Botanical Research" 1:209-278, Academic Press, London, 1963

32. Lambers, H, Efficiency of root respiration in relation to growth rate, morphology and soil composition. Physiol Plant 46:194-202, 1979

33. Lambers, H and SJ Van Dijk, Cyanide-resistant root respiration and tap root formation in two subspecies of *Hypochaeris radicata*. Physiol Plant 45: 235-239, 1979

34. Lawn, RJ and WA Brun, Symbiotic nitrogen fixation in soybeans. I. Effect of photosynthetic source sink manipulations. Crop Sci 14:11-16, 1974

35. Layzell, DB, RM Rainbird, CA Atkins and JS Pate, Economy of photosynthate use in nitrogen-fixing legume nodules: observations on two contrasting symbioses. Plant Physiol 64:888-891, 1979

36. Layzell, DB, JS Pate, CA Atkins and DT Canvin, Partitioning of carbon and nitrogen and the nutrition of root and shoot apex in a nodulated legume. Plant Physiol (in press), 1981

37. Lewis, DAM and JS Pate,.The significance of transpirationally derived nitrogen in protein synthesis in fruiting plants of pea (*Pisum sativum* L). J Exp Bot 24:596-606, 1973

38. MacRobbie, EAC, Phloem translocation. Biol Rev 46:429-81, 1971

39. Mahon, JD, Respiration and the energy requirement of nitrogen fixation in nodulated pea roots. Plant Physiol 60:817-821, 1977

40. Marshall, C and GR Sagar, The influence of defoliation on the distribution of assimilates in *Lolium multiflorum* Lam. Ann Bot, London, 29:365-370, 1965

41. McKee, HS, Nitrogen Metabolism in Plants. Clarendon, Oxford, 1962

42. McNeil, DL, CA Atkins and JS Pate, Uptake and utilization of xylem-borne amino compounds by shoot organs of a legume. Plant Physiol 63:1076-81, 1979

43. Metcalfe, CR and L Chalk, Anatomy of the Dicotyledons. Clarendon, Oxford, 1950

44. Miflin, BJ and PJ Lea, Amino acid metabolism. Ann Rev Plant Physiol 28:299-329, 1977

45. Milburn, JA, Pressure Flow, In: Encyclopedia of Plant Physiology I. Transport in Plants, I. Phloem Transport (Zimmermann, MR and JA Milburn eds). Springer, Berlin, 1975

46. Oghoghorie, CGO and JS Pate, Exploration of the nitrogen transport system of a nodulated legume using ^{15}N. Planta 104: 35-49, 1972

47. Pate, JS, Root exudation studies on the exchange of C^{14}-labelled organic substances between the roots and shoot of the nodulated legume. Plant and Soil 17:333-356, 1962

48. Pate, JS, Uptake, assimilation and transport of nitrogen compounds by plants. Soil Biol Biochem 5:109-119, 1973

49. Pate, JS, Exchange of solutes between phloem and xylem and circulation in the whole plant, In: Encyclopedia of Plant Physiology I. Transport in Plants, I. Phloem Transport. (Zimmermann, MH and JA Milburn eds) Springer, Berlin, 1975

50. Pate, JS, Transport and partitioning of nitrogenous solutes. Ann Rev Plant Physiol 31:313-340, 1980

51. Pate, JS, CA Atkins, K Hamel, DL McNeil and DB Layzell, Transport of organic solutes in phloem and xylem of a nodulated legume. Plant Physiol 63:1082-1088, 1979

52. Pate, JS, CA Atkins, DF Herridge and DB Layzell, Synthesis, storage and utilization of amino compounds in white lupin (*Lupinus albus* L). Plant Physiol (in press), 1981

53. Pate, JS, CA Atkins, ST White, RM Rainbird and KC Woo, Nitrogen nutrition and xylem transport of nitrogen in ureide-producing grain legumes. Plant Physiol (in press), 1980

54. Pate, JS and AM Flinn, Fruit and Seed Development, In: "The Physiology of the Garden Pea" (Sutcliffe, JP and JS Pate eds). Academic Press, London pp 441-468, 1977

55. Pate, JS and DF Herridge, Partitioning and utilization of net photosynthate in a nodulated annual legume. J Exp Bot 29:401-412, 1978

56. Pate, JS, DB Layzell and DL McNeil, Modelling the transport and utilization of carbon and nitrogen in a nodulated legume. Plant Physiol 63:730-737, 1979

57. Pate, JS and FR Minchin, Comparative studies of carbon and nitrogen of selected grain legumes, In: Advances in Legume Science (Summerfield, RJ and AH Bunting eds), Royal Botanic Gardens, Kew, England, 1980

58. Pate, JS and TP O'Brien, Micro-autoradiographic study of the incorporation of labelled amino acids into insoluble compounds of the shoot of a higher plant. Planta (Berl) 78:60-71, 1968

59. Pate, JS, PJ Sharkey and CA Atkins, Nutrition of a developing legume fruit. Functional economy in terms of carbon, nitrogen and water. Plant Physiol 50:506-510, 1977

60. Peel, AJ, Transport of Nutrients in Plants. Butterworths, London, 1974

61. Proctor, JTA, RL Watson and JJ Landsberg, The carbon budget of a young apple tree. J Am Soc Hort Sci 101:579-583, 1976

62. Quinlan, JD and GR Sagar, An autoradiographic study of the movement of ^{14}C-labelled assimilates in the developing wheat plant. Weed Research 2:264-273, 1962

63. Radin, JW, Contribution of the root system to nitrate assimilation in whole cotton plants. Aust J Plant Physiol 4:811-819, 1977

64. Raper, CD, LR Parsons, DT Patterson and PJ Kramer, Relationship between growth and nitrogen accumulation for vegetative cotton and soybean plants. Bot Gaz 138:129-137, 1977

65. Raper, CD, DT Patterson, LR Parsons and PJ Kramer, Relative growth and nutrient accumulation rates for tobacco. Plant and Soil 46:473-486, 1977

66. Rawson, HM and C. Hackett, An exploration of the carbon economy of the tobacco plant. III Gas exchange of leaves in relation to position on the stem, ontogeny and nitrogen content. Aust J Plant Physiol 1:551-560, 1974

67. Ryle, GJA, CE Powell and AJ Gordon, Effect of source of nitrogen on the growth of Fiskeby soya bean: the C economy of whole plants. Ann Bot 42:637-648, 1978

68. Ryle, RJA, CE Powell and AJ Gordon, Respiratory costs of N_2 fixation in soyabean, cowpea and white clover. 2. Comparisons of the cost of N_2 fixation and the utilization of combined N. J Exp Bot 30:145-155, 1979

69. Schrader, LE, Uptake, accumulation and transport of nitrogen in higher plants. Nitrogen in the Environment 2:101-141, 1978

70. Stiles, W and W Leach, Respiration in Plants. Methuen & Co Ltd, London, 1952

71. Sutcliffe, JF and ODG Collins, A mechanism of phloem and transport based on interfacial flow controlled by solute potential gradients. Ann Bot 39:627-9, 1975

72. Thomas, RJ, U Feller and KH Erismann, The effect of different inorganic nitrogen sources and plant age on the composition of bleeding sap of *Phaseolus vulgaris*. New Phytol 82:657-70, 1979

73. Turgeon, R and JA Webb, Leaf development and phloem transport in *Curcubita pepo:* carbon economy. Planta 123:53-62, 1975

74. Wardlaw, IF, Phloem transport: physical, chemical or impossible. Ann Rev Plant Physiol 25:515-539, 1974

75. Wieringa, KT and JA Bakhuis, Chromatography as a means of selecting effective strains of *Rhizobia*. Plant and Soil 8:254-260, 1957

76. Williams, JH, The physiology of groundnuts (*Arachis hypogaea* L cv Egret) 2. N accumulation and distribution. Rhod J Agric Res 17:49-56, 1979

77. Wych, RD and DW Rains, Nitrate absorption and acetylene reduction by soybeans during reproductive development. Physiol Plant 47:200, 1979

78. Ziegler, H, Nature of transported substances, In: Encyclopedia of Plant Physiology I. Transport in Plants, I. Phloem Transport (Zimmermann, MR and JA Milburn eds). Springer, Berlin, 1975

79. Zimmermann, MH, Translocation of nutrients, In: "The Physiology of Plant Growth and Development" (Wilkins, MB ed). McGraw Hill, Maidenhead, England, 1969

CHAPTER 5

RELATIONSHIPS BETWEEN NITROGEN METABOLISM AND PHOTOSYNTHESIS

James A. Bassham, Peder O. Larsen*,
Arthur L. Lawyer and Karen L. Cornwell

Laboratory of Chemical Biodynamics, Lawrence Berkeley Laboratory, University of California, Berkeley CA 94720 USA
and
*Chemistry Department, Royal Veterinary and Agricultural University, 40 Thorvaldsensvej - DK 1871 Copenhagen V. Denmark

Introduction
Chloroplast Synthesis of Keto Acids and Amino acids
Bidirectionality of Glycine-Serine Interconversion
Regulation of Carbon Flow from Photosynthesis to Amino Acid Biosynthesis
Summary
Acknowledgement
References

J.D. Bewley (ed.), Nitrogen and Carbon Metabolism

INTRODUCTION

Photosynthetic green cells generate reducing power from the oxidation of water to O_2, and use the reducing power for the reduction of CO_2, nitrate and sulfate. The principal products of green cells then are oxygen, sucrose and other carbon compounds, amino groups of amino acids, and sulfhydryl groups of amino acids.

Even if the biosynthetic pathways leading to these various organic compounds were separate after the initial act of reduction, regulation would be required to allocate the reducing power and appropriate organic precursors to the required biosynthetic pathways. Because of photorespiration and its consequent required carbon pathways for the recovery of reduced carbon, flows of nitrogen and carbon are further intertwined at the conversions of glycolate, glyoxylate, glycine, glutamate, serine, and hydroxypyruvate. The glycolate pathway of photorespiration operates between the chloroplasts, peroxisomes, mitochondria, and cytoplasm. Thus, various transport mechanisms are involved, and a careful regulation of these transport mechanisms is required.

There is also the question of biosynthetic capabilities of the various organelles and compartments in green cells. For example, which keto acid skeletons can the chloroplast make for itself from CO_2 for amino acid synthesis, and which must be imported?

These issues are being addressed by papers that precede and follow this one, so that considerable overlap is expected. At the same time, the range of issues is so broad that it is doubtful that one person could hope to deal adequately with them all. This discussion will be limited, therefore, to three issues about which investigators in our laboratory can provide some information. These issues are first, the keto acid and amino acid biosynthetic capabilities of the chloroplast; second, the bidirectionality of the glycine-serine interconversion associated with photorespiration or its absence; and third, the complex issue of the regulation of carbon flow from photosynthesis to biosynthesis of starch and sucrose as compared with synthesis of amino acids in the photosynthetic cell. In particular, we will discuss the profound regulatory effects of the addition of low levels of ammonium ion to cell media. These were seen first with Chlorella pyrenoidosa, and then with spinach leaf discs, poppy cells and now with isolated spinach leaf cells.

CHLOROPLAST SYNTHESIS OF KETO ACIDS AND AMINO ACIDS

The pathway of nitrogen assimilation in plants and especially in chloroplasts, has been reviewed recently by Lea and Miflin (1), who in 1974 first discovered the key enzyme in glutamate synthesis in chloroplasts (2). This enzyme, glutamine-oxyglutarate aminotransferase (GOGAT) utilizes reduced ferredoxin from the light reactions in the chloroplast to bring about the conversion of one molecule of oxoglutarate and one molecule of glutamine to two molecules of glutamate. In green cells this reaction appears to occur mostly, if not exclusively, in the chloroplast. The synthesis of glutamine, requiring ATP, ammonia, and glutamic acid and mediated by glutamine synthetase, occurs partly in the chloroplast and partly in the cytoplasm. In pea cells about half of the leaf glutamine synthetase is located in the chloroplast with the remainder in the cytoplasm (3).

The synthesis of glutamate in the chloroplast may occur in the green cell on a scale far in excess of the needs of biosynthetic pathways for glutamate. Keys et al. (4)

proposed that there is a photorespiratory nitrogen cycle required to recapture the ammonia liberated by the conversion of two glycine molecules to one serine molecule in that part of the glycolate pathway which occurs in the mitochondria. The ammonia is recaptured by conversion of glutamate to glutamine with ATP in the glutamine synthetase reaction. Keys et al. believe that this ammonia incorporation occurs in the cytoplasm since it would be there that ammonia first encounters glutamine synthetase activity. The glutamate for glutamine synthesis is formed by the action of GOGAT in the chloroplast (Fig. 1). The other glutamate formed by the GOGAT reaction is utilized in the peroxisome to provide an amino group for transamination of glyoxylate to glycine, releasing oxoglutarate in the process which is then recycled to the chloroplasts for the GOGAT reaction.

Although the major amount of glutamine synthesis may be occurring via the photorespiratory nitrogen cycle, any net synthesis of glutamate for subsequent biosynthesis, including protein synthesis, requires that there be a net conversion of oxoglutarate to glutamate. Presumably oxoglutarate is derived from anaplerotic reactions occurring in the mitochondria. As we shall discuss later, the flow of carbon from photosynthesis into the TCA cycle in the mitochondria can be stimulated by the addition of small amounts of ammonia external to the photosynthetic cell. In the light both the conversion of phosphoenylpyruvic acid (PEP) to pyruvate and the carboxylation of PEP to give oxaloacetate are increased in the presence of added ammonium ion. Consequently, an increased supply of ketoacid skeletons becomes available in the form of oxoglutarate for the synthesis of glutamate.

Mitchell and Stocking (5) have shown that isolated chloroplasts can assimilate ammonia into glutamine when provided with glutamate. Conversion of oxoglutarate to glutamate primarily occurs via GOGAT while ammonia incorporation occurs via glutamine synthetase (1). Glutamate dehydrogenases are present in plant cell chloroplasts and mitochondria, but are considered to be involved only in degradative reactions because of their high K_m values for ammonia (see ref. 1, for example).

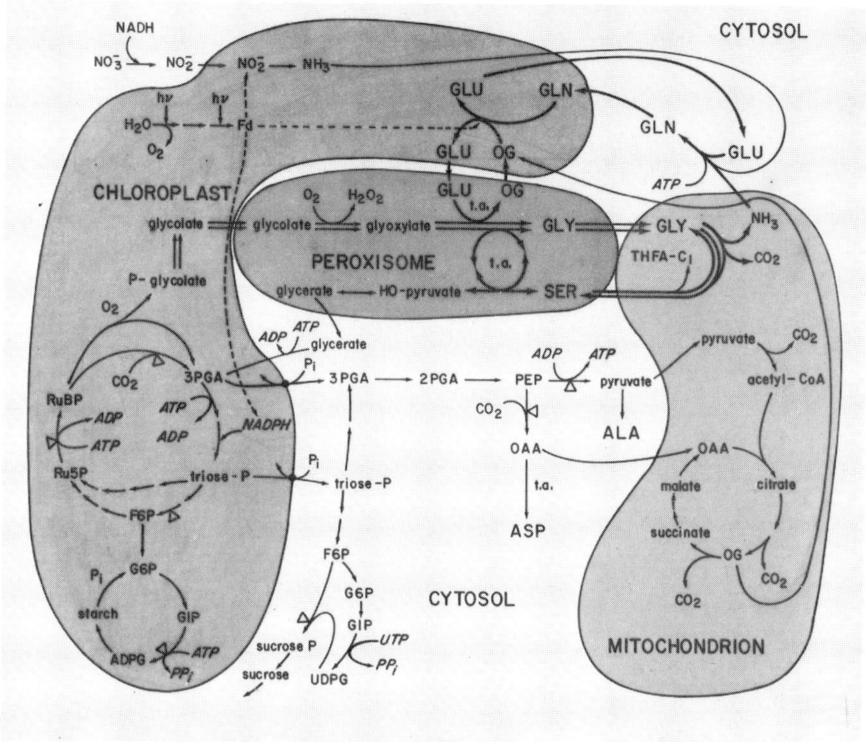

Figure 1. Carbon and nitrogen photosynthetic and photorespiration pathways. Abbreviations: RuBP, ribulose-1,5-bisphosphate; Ru5P, ribulose-5-phosphate; PGA, 3-phosphoglycerate; F6P, fructose-6-phosphate; G6P, glucose-6-phosphate; G1P, glucose-1-phosphate; ADPG, adenosine diphosphoglucose; UDPG, uridine diphosphoglucose; PEP, phosphoenolpyruvic acid; OG, 2-oxoglutaric acid; standard amino acid abbreviations. The reactions of the photorespiratory nitrogen cycle (adapted from Keys et al., ref. 4) are indictaed by heavier lines.

It is generally accepted that net synthesis of glutamate and glutamine in chloroplasts depends on five-carbon ketoacid skeletons derived from the mitochondria. Similarly, there is abundant evidence that in both C_4 and C_3 plants, oxaloacetate formed by PEP carboxylation in the cytoplasm can be utilized in the chloroplast for the synthesis of aspartic acid and asparagine. These important four- and five-carbon amino acids can then serve as substrates for the synthesis of a variety of other amino acids in the chloroplasts. Given these possibilities, what is the relative importance of the biosynthesis of amino acids in the chloroplasts as compared with that in the green cell as a whole? There has been much discussion of the biochemical autonomy of green cell chloroplasts (see for example review by Givan and Leach (6)). The generally accepted view appears to be that chloroplasts can form most if not all required amino acids from the appropriate ketoacid skeletons, but that some of these ketoacid skeletons may have to be imported from other parts of the cell in a manner similar to that just discussed for four- and five-carbon skeletons. There has been some interest, therefore, in the capability of isolated leaf cell chloroplasts to carry out the synthesis of amino acids from $^{14}CO_2$ during photosynthesis.

Relatively crude chloroplast preparations such as those used by Jensen and Bassham (7) produced very little labeled amino acids compared to the major products of $^{14}CO_2$ photosynthesis. This absence or near absence of appreciable amounts of labeled amino acids from photosynthesizing crude chloroplast preparations has led us in the past to believe that amino acid biosynthesis in chloroplasts is indeed heavily dependent on external supply of three-, four-, and five-carbon ketoacid skeletons. Kirk and Leach (8), investigating amino acid biosynthesis by isolated chloroplasts during photosynthesis, found significant labeling of aspartate, alanine, glycine, serine, threonine and lysine, with the greater part of the label being in aspartate. More recently, Buchholz et al. (9) investigated amino acid biosynthesis during isolated chloroplast photosynthesis using unpurified intact chloroplasts as well as chloroplasts purified by the method of Larsson and Albertson (10). These authors found considerably more incorporation of ^{14}C into amino acids in their "unpurified" chloroplasts than we normally do, but the amount of incorporation of ^{14}C into amino acids was reduced to very low levels in the purified chloroplast. Of that found, the largest fixation was into alanine with lesser amounts into the four- and five-carbon amino acids, and serine and glycine. Addition of leaf perioxsomes restored glycine synthesis and addition of leaf mitochondria resulted in serine synthesis. On the other hand, the rate of aromatic amino acid synthesis in the purified chloroplast was not enhanced by the addition of other organelles. This supports the proposition that the shikimic acid pathway of biosynthesis of aromatic compounds is mainly located in the chloroplast (11).

Recently we reinvestigated the question of isolated chloroplast competence for amino acid biosynthesis. The results are preliminary and generally confirm existing views as just stated, but there are several interesting aspects to the recent results. The spinach chloroplasts were prepared according to the Jensen and Bassham method (7) but were washed twice with buffer solution to remove most cytoplasmic impurities. Nevertheless, this procedure leaves the chloroplast contaminated with peroxisomes. This system, while not representing the situation of pure chloroplasts, was interesting particularly when the experiment was carried out as it was with the steady-state apparatus (12) which provided for $^{14}CO_2$ fixation at air levels of CO_2 and O_2. Parallel experiments were carried out with isolated whole spinach cells also photosynthesizing with air levels of CO_2 and O_2. Various additions of nitrite, oxoglutarate, glutamate and glutamine either alone or in combination were made to the isolated chloroplasts. The details of the investigations will be published elsewhere, but a few points are worth mentioning.

The chloroplasts contaminated with peroxisomes incorporated about 20% as much ^{14}C into glycine as did the whole cells (Table 1). Addition of 0.1 mM glutamate increased glycine formation in the chloroplast preparations to 140% of that of whole cells, while addition of 0.1 mM oxoglutarate decreased it to 13%. Addition of both oxoglutarate and glutamine, which would be expected to produce glutamate by the GOGAT reaction, increased glycine formation to 31%. All of these effects are confirmatory of the scheme of nitrogen cycling shown in Fig. 1. It is interesting that this much glycine synthesis can be accomplished by isolated chloroplasts contaminated with peroxisomes. One might expect considerable dilution of the glycolate produced by photorespiration if it goes into the suspending medium before finding its way into the peroxisomes. Peroxisomes in green leaf cells are, in fact, frequently seen to be tightly appressed to chloroplasts (see, for example, Frederick and Newcomb (13)). It certainly is possible that in leaf cells there could be a direct flow of glycolate from chloroplasts into peroxisomes.

Table 1. Incorporation of ^{14}C into amino acids in isolated spinach Chloroplasts (+ peroxisome contamination) compared with whole isolated spinach cells (expressed as percent).

Medium	Glycine	Alanine	Serine	Valine	Phenylalanine
Control	20	2.8	3.5	110	18
+ glutamate	140	7.6	3.3	90	29
+ oxoglutarate	13	1.2	0.4	--	14
+ glutamine	25	6.4	1.3	120	5.4
+ NO_2^-	11	3.9	1.4	60	6.0
+ oxoglutarate + glutamine	31	2.8	2.3	--	24

Chloroplasts were isolated by procedure of Jensen and Bassham (7). Chloroplast photosynthesis with $^{14}CO_2$ was for 40 min. Spinach cells were isolated by a modification of method of Paul and Bassham (24) and photosynthesized for 30 min (Larsen et al., to be published).

The amount of ^{14}C incorporated into most other major amino acids in the chloroplast preparations was in the range of a few percent compared with whole cell incorporation. Alanine and serine serve as examples of this incorporation (Table 1). Whether these low amounts of biosynthesis from $^{14}CO_2$ of common amino acids by isolated chloroplasts represent chloroplastic synthesis or synthesis due to contamination cannot be said from these experiments, but there is other evidence in the literature to suggest that small amounts of these amino acids can in fact be synthesized by chloroplasts (8). In contrast to these small amounts are the relatively large amounts of valine and phenylalanine found with the crude preparations of chloroplasts. As already mentioned, phenylalanine may be labeled by synthesis via the shikimic acid pathway which occurs in chloroplasts.

The synthesis of almost the entire cellular requirement of valine by isolated chloroplasts (Table 1) provides further strong evidence for the presence in chloroplasts of adequate levels of pyruvate and "active acetaldehyde" derived from pyruvate since these are the precursors in valine biosynthesis. The fact that only a small part of the cell's alanine synthesis occurs in the chloroplasts must therefore be due to the presence of active synthesis of alanine outside the chloroplast rather than to the absence of pyruvate in the chloroplast.

Addition of oxoglutarate had a generally negative effect on synthesis of amino acids by the isolated chloroplasts, presumably by shifting the equilibrium of glutamate aminotransferases in the direction of glutamate formation. By the same token, added glutamate greatly increased the synthesis of some amino acids, notably glycine.

In summary, it appears that chloroplasts do indeed have a nearly complete capability for the synthesis of all amino acids, provided only that there is an external supply of oxoglutarate and oxaloacetate. The chloroplasts do apparently play a prominent role in the synthesis of certain amino acids for the cell. These include amino acids such as phenylalanine and tyrosine that are formed by the shikimic acid pathway and valine which is formed from pyruvate. Given the apparently intimate association between peroxisomes and chloroplasts, it would not be surprising if the chloroplasts obtained their complement of glycine from the peroxisomes, but the possibility of chloroplast synthesis of small amounts of glycine cannot be excluded. It should be noted that, in general, developed green leaf cells devote only a relatively small proportion of their total photosynthate to the synthesis of amino acids for the purpose of subsequent protein, pigment, and nucleotide biosynthesis. The major amino acid biosynthetic activity of green cells has to do with the binding of the ammonia which is formed by the reduction of nitrate to nitrite in the cytoplasm and nitrite to ammonia in the chloroplast. This binding results in the synthesis of glutamate, glutamine, and perhaps asparagine. Amino-containing compounds are then translocated to other organs of the plant.

BIDIRECTIONALITY OF GLYCINE-SERINE INTERCONVERSION

Formation of glycine and serine in C_3 plants is generally attributed to the glycolate pathway sequence: glycolate to glycine to serine to glycerate, believed to result from the process of photorespiration as shown in Fig. 1 (14,15). An alternative pathway is the reverse sequence from 3-phosphoglycerate (PGA) to glycerate to serine to glycine (16). The glycolate to glycerate pathway is expected under conditions of photorespiration such as exist with C_3 plants in air with bright illumination and warm temperatures. The reverse pathway is generally expected to occur under conditions where there is little or no photorespiration such as might exist with elevated CO_2 concentration, for example.

Platt et al. (17) found that in alfalfa leaflets photosynthesizing under steady-state conditions, elevated concentrations of $^{14}CO_2$ greatly decreased the amount of label in glycine and glycolate, as would be expected with diminished photorespiration. They also found, however, that pool sizes of serine and glycerate reflected the size of the PGA pool under steady state conditions with either air levels or high levels of CO_2. While recognizing that these results may be consistent with the reversibility of the pathway, they suggested that the parallel between serine and PGA pool sizes under both conditions might indicate a predominent formation of serine from PGA even with air levels.

In view of this and other uncertainties with respect to the magnitude and direction of flow between serine and glycine, it was fortunate that Lawyer and Zelitch (18) found that with tobacco callus tissue, glycine hydroxamate acts as a rather specific inhibitor of the conversion of glycine to serine and CO_2 in the glycolate pathway. This inhibitor has now been used by Lawyer et al. (to be published) to investigate the bidirectionality of flow between glycine and serine in isolated spinach cells photosynthesizing under steady-state conditions with several levels of O_2 and CO_2. For comparison, experiments were also carried out with 7 mM bicarbonate (saturating CO_2). As expected (Fig. 2), the highest rates of spinach cell photosynthesis were obtained with high CO_2 and low O_2 as well as with bicarbonate in air. Nearly as high rates were obtained with high CO_2 and high O_2, although there was a lag at the beginning of the fixation period. Much lower rates of photosynthesis were seen with air level CO_2 and with low O_2. Similarly, low rates of photosynthesis were seen with air level CO_2 and with high O_2, and in this case there was again a lag at the beginning of the fixation period. Thus the effect of CO_2 pressure is on the long-term rate of photosynthesis whereas the effect of oxygen pressure appears mainly in the time required to achieve the full rate of steady-state photosynthesis.

In parallel with each of the experiments shown in Fig. 2, duplicate flasks contained spinach cells to which up to 3.8 mM glycine hydroxamate had been added. In all cases the cells were killed and analyzed for levels of amino acids and other metabolites by two-dimensional paper chromatography following a preliminary ion exchange separation of neutral, acidic, and basic amino acids from other components. As expected (Table 2), the level of labeled glycine was highest in the control with low CO_2 and high O_2 (air, photorespiratory conditions), and this high level was further increased in the flask with added glycine hydroxamate. In both intermediate cases (low CO_2, low O_2 and high CO_2, high O_2) addition of glycine hydroxamate also increased the level of glycine indicating flow in the direction from glycine to serine. Under the condition where the least photorespiration would be expected (high CO_2, low O_2) addition of glycine hydroxamate decreased the level of labeled glycine by 73%.

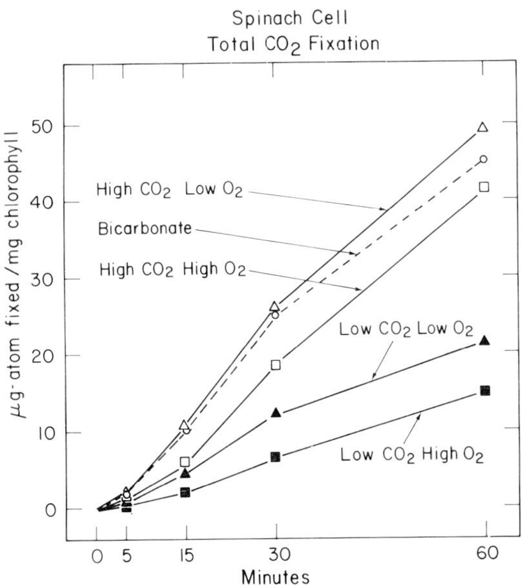

Figure 2. Total $^{14}CO_2$ fixation into isolated spinach mesophyll cells at 22°C and 580 μE·m^{-1}·sec^{-1} under the following conditions: △, 0.21% CO_2 and 2.3% O_2; □, 0.04% CO_2 and 2.4% O_2; ▲, 0.20% CO_2 and 20.0% O_2; ■, 0.038% CO_2 and 20.2% O_2; ○, 8 mM $H^{14}CO_3^-$.

In this case, where we would expect glycine to be formed from serine, it appears that glycine hydroxamate inhibits serine to glycine conversion. Not all steps in the serine-glycine interconversions are truly reversible, but it appears that some key step in both directions is inhibited by glycine hydroxamate. In the two intermediate cases (low CO_2, low O_2, and high CO_2, high O_2) despite very different overall photosynthetic rates (Fig. 1), addition of glycine hydroxamate increased the level of glycine by about 60% suggesting that in these cases, the net flow of carbon was from glycine to serine.

Table 2. ^{14}C labeling of glycine in isolated spinach cells after 1 h photosynthesis with $^{14}CO_2$ - effect of glycine hydroxamate

	μg-atoms $^{14}C \cdot mg\ Chl^{-1} \cdot hr^{-1}$			
	High CO_2 Low O_2	Low CO_2 Low O_2	High CO_2 High O_2	Low CO_2 High O_2
Control	0.12	0.12	0.061	0.29
+ glycine hydroxamate	0.032	0.20	0.097	0.35

Isolated spinach mesophyll cells were allowed to photosynthesize with $^{14}CO_2$ with or without 3.8 mM glycine hydroxamate in the medium at 22 C and 580 $\mu E \cdot m^{-2} \cdot sec^{-1}$. The four gas mixtures were: 0.20% CO_2 and 2.3% O_2; 0.035% CO_2 and 2.6% O_2; 0.19% CO_2 and 18.8% O_2; and 0.044% CO_2 and 20% O_2. Total CO_2 fixation rates (μg-atoms·mg $Chl^{-1} \cdot hr^{-1}$) were 20.1 and 23.5; 15.4 and 15.3; 26.0 and 21.1, and 16.9 and 9.9 without and with glycine hydroxamate, respectively.

The changes in serine induced by the inhibitor are more complex (Table 3). Under conditions of low photorespiration, the level of serine was unchanged. Under all of the other three conditions, the level of serine was increased significantly. The explanation for this lies in the fact that the glycine hydroxamate, in blocking the flow of carbon from glycine to serine, also blocks the release of ammonia. As a consequence of this blockage, glycine increases dramatically as already mentioned. This increase in glycine concentration leads to a large shift in the equilibrium mediated by serine-glyoxylate aminotransferase in the direction of serine and glyoxylate formation. This glyoxylate plus new glyoxylate coming from photorespiration uses up more glutamic acid to make more glycine and oxoglutaric acid so that the net result is the conversion of amino groups from glutamate to the pools of serine and glycine. As a consequence of the decreased glutamate pool, considerably less fixation of carbon is seen in amino acids other than glycine and serine.

Table 3. *^{14}C labeling of serine in isolated spinach cells after 1 h photosynthesis with $^{14}CO_2$ - effect of glycine hydroxamate*

	µg atoms $^{14}C \cdot mg\ Chl^{-1} \cdot hr^{-1}$			
	High CO_2 Low O_2	Low CO_2 Low O_2	High CO_2 High O_2	Low CO_2 High O_2
Control	0.15	0.21	0.17	0.32
+ glycine hydroxamate	0.13	0.57	0.39	0.98

Conditions were the same as described for Table 2.

Apparently, the pathway from PGA to glycerate to hydroxypyruvate can supply carbon to serine under these conditions when the flow of carbon from glycine to serine is blocked. The level of glycerate was highest in the photorespiratory condition (Table 4, low CO_2 and high O_2), and was decreased 60% by the presence of glycine hydroxamate. In the intermediate conditions glycerate levels were lower, but nonetheless were decreased upon addition of glycine hydroxamate, presumably due to conversion to hydroxypyruvate which in turn is converted to serine as a consequence of glycine buildup.

Table 4. ^{14}C labeling of glycerate in isolated spinach cells after 1 hr photosynthesis with $^{14}CO_2$ - effect of glycine hydroxamate

	μg atoms $^{14}C \cdot mg\ Chl^{-1} \cdot hr^{-1}$			
	High CO_2 Low O_2	Low CO_2 Low O_2	High CO_2 High O_2	Low CO_2 High O_2
Control	0.070	0.066	0.097	0.136
+ glycine hydroxamate	0.058	0.056	0.048	0.027

Conditions were the same as described for Table 2.

These results add further support for the already well-accepted pathway of glycine to serine conversion during conditions of photorespiration. They also support the existence of the reverse pathway from serine to glycine under conditions where photorespiratory formation of glycine is minimal. In terms of overall steady state photosynthesis rate, CO_2 concentration appears to be considerably more important than O_2 concentration in the ranges tested. With 0.2% CO_2, the rate of photosynthesis during the second half hour was as high with 20% O_2 as it was with 2% O_2 despite the fact that inhibitor data indicate a photorespiratory formation of glycine at 20% O_2 though at a much lower level than with air. Even with air levels of CO_2, the long-term rate of photosynthesis is unaffected by the difference between 20% O_2 and 2% O_2 at least with these isolated spinach cell preparations. Conceivably, cells exhibiting higher photosynthetic rates could show a greater dependency on oxygen concentration. In any event, the present experiments clearly demonstrate the bidirectionality of glycine-serine interconversion.

REGULATION OF CARBON FLOW FROM PHOTOSYNTHESIS TO AMINO ACID BIOSYNTHESIS

Since mature leaf cells are primarily devoted to the synthesis of sucrose for export to the rest of the plant, the direct synthesis of amino acids and other cellular constituents is normally limited in amount, though of great importance to the cell's metabolism. Because of this and other reasons, it is somewhat easier to examine regulation of photosynthetic carbon metabolism in unicellular algae which are undergoing a complete cell cycling through growth and division. Ten years ago, Kanazawa et al. (19) established the pattern of change in secondary metabolism in synchronously growing Chlorella pyrenoidosa during the cycling of cells from recently divided cells through the growth stage, predivision and division. Profound shifts in the allocation of carbon to sucrose, proteins and amino acids and nucleic acids were seen over the course of the cell cycle. Accompanying the shifts were changes in the rate of reduction of nitrate to amino acids (20). Since these changes in the rate of reduction of nitrates to ammonia resulted in changes in supplies of ammonia, the effects of ammonia itself on the regulation of carbon metabolism were investigated (21,22).

Not surprisingly, the addition of small amounts of ammonia (1 mM) to the medium of the photosynthesizing Chlorella resulted in considerable stimulation of the synthesis of certain amino acids, especially glutamine, glutamic acid, alanine, and aspartate. Further investigation revealed that this stimulation was not the result solely of an increased supply of ammonia for amino groups, but was in fact a consequence of an accelerated flow of carbon from photosynthesis into intermediates of the TCA cycle such as citrate with a concurrent decrease in the flow of carbon into starch and especially sucrose. Besides the increased flow of carbon into amino acids, there was an increased rate of flow of carbon from photosynthesis into the synthesis of fatty materials. A principal cause of this latter increase was found to be a stimulation of the conversion of phosphoenolpyruvic acid (PEP) to pyruvic acid, presumably through a stimulation of the activity of pyruvate kinase. Since increased synthesis of amino acids derived from the TCA cycle requires anaplerotic reactions, it was assumed that carboxylation of PEP to produce oxaloacetic acid also increased (21,22).

Investigation of the effects of adding ammonia to respiring Chlorella pyrenoidosa (in the dark following a period of photosynthesis to label the carbon intermediates) showed that the effects of ammonia could be seen equally well in the dark and were generally interpretable as effects occurring outside of the chloroplast and in the cytoplasm. When the light was turned off following a period of photosynthesis by Chlorella which had been starved for nitrogen, the formation of sucrose stopped, but sucrose was not utilized for respiration, whereas the store of starch, presumably in the chloroplast, became utilized for respiratory reactions as indicated by a steady decline in the level of labeled material. When the light was turned off, there was a temporary drop in the level of ATP, but the level was then restored to the same level in the dark as in the light, presumably by the operation of oxidative phosphorylation. The light to dark change did not in this condition result in any increase in the flow of carbon into amino acid biosynthesis (22).

When ammonia was subsequently added to the medium in the dark, the rate of starch utilization remained constant and the level of sucrose began immediately to decline and continued to do so at a constant rate. Thus, both sucrose and starch were now contributing carbon to respiratory and biosynthetic metabolism in the cells. Also, with the addition of ammonium ion, the steady-state level of ATP declined momentarily and then rose again but to a lower steady-state level than before. This is surprising because ammonium ion, while known to have an effect on photophosphorylation in the chloroplast, would not be expected to affect oxidative phosphorylation in the mitochondria. With the addition of ammonia to the respiring *Chlorella*, the steady-state level of PEP dropped and the steady state level of pyruvate rose, clearly indicating an acceleration of the pyruvate kinase reaction. Other increases were seen in the labeling of citrate and the amino acids, glutamate, glutamine, alanine and some others. Thus, it was clear that the presence of 1 mM ammonium ion in the medium of the *Chlorella* had produced a profound shift in metabolism from sucrose to amino acid and fatty acid biosynthesis in the light and an accelerated utilization of sucrose in the dark accompanied by an increased amino acid synthesis. The regulatory mechanism for pyruvate kinase is not known but could be related to the change in the levels of ATP (22).

Although the primary photosynthetic reactions and their regulation and chloroplast metabolism may be rather similar between *Chlorella pyrenoidosa* and green cells of higher plants, there is no reason to think that subsequent biosynthetic pathways and their regulation in the cytoplasm and plant cell organelles would be the same, particularly considering the fact that unicellular algae apparently do not exhibit the same glycolate pathway for conversion of glycine to serine with its consequent large release of ammonia and recycling of ammonia within the cell. In the case of the unicellular algae *Chlorella*, it did appear that the reduction of nitrate to nitrite and the nitrite to ammonia, and the consequent rate of ammonia formation might play a rather significant role in the overall regulation of carbon metabolism. It was of interest to see what effect, if any, small amounts of ammonium ion added to leaves or cells of leaves in higher plants might have on the regulation of carbon metabolism in such cells.

When alfalfa leaf discs were floated on a buffer containing ammonium chloride (5 mM), photosynthesis with

$^{14}CO_2$ produced more labeled amino acid and less sucrose than in the control disc without ammonia (23). The level of pyruvate increased and that of PEP decreased. These and other changes in labeling of secondary metabolites led to the conclusion that pyruvate kinase was activated in the presence of ammonia resulting in an increased transfer of photosynthetically incorporated carbon to synthesis of amino acid skeletons at the expense of sucrose synthesis.

Given the limitations of leaf discs in terms of uniform penetration of added chemicals to all the cells, it became desirable to carry out experiments with isolated leaf cells. Methods were developed for maintaining cells isolated from poppy plants (Papaver somniferum L.)(24) and these cells were used in studies with added 1 mM and 2 mM ammonium ion (25). Addition of ammonia to leaf cells caused an immediate rapid increase in the level of pyruvate and of alanine which is formed from pyruvate by transamination. The ratio pyruvate:PEP rose immediately and rapidly upon the addition of ammonia, clearly indicating the stimulation of the flow of carbon via the pyruvate kinase reaction. At the same time, there was an even more dramatic sudden increase in the ratio of aspartate:PEP. Since the levels of labeled oxoglutarate and glutamate were also measured, it was possible to calculate, via the oxaloacetate-glutamate transamination reaction, that oxaloacetate labeling had rapidly increased upon addition of ammonia, and that consequently the PEP carboxylase step had been dramatically stimulated by the addition of the ammonium ion.

The increased flow of carbon from photosynthesis into the TCA cycle could also be seen by a large stimulation in the rates of labeling of both malate and citrate. The level of glutamate itself dropped momentarily, presumably due to the stimulation of glutamine formation from glutamate by the added ammonia, but then rose more rapidly than the control despite the fact that there was a general stimulation of amino acid synthesis which would be using up glutamate via the transaminase catalyzed reaction. After 25 min with cells in the presence of ammonia, the sucrose pool was about 16% less than in the control. Starch synthesis in this case was largely unaffected.

From these results it may be concluded that in these cells isolated from leaves of higher plants, as in the case of the *Chlorella*, the addition of ammonium ion resulted in a shift in metabolism from the formation of sucrose to the formation of increased amounts of amino acids and that this shift was produced not only by the availability of ammonia for amino group formation but by also regulatory phenomenon resulting in increased flow of carbon from photosynthetic products to the TCA cycle. That the magnitude of the shift away from sucrose synthesis was not as great in this case of leaf cells might be attributed to the less complete nitrogen starvation prior to the beginning of the experiment. Another difference from the effects on *Chlorella* was noted when the studies with the poppy cells were extended to addition of ammonium in the dark following a period of photosynthesis (26). The stimulation of PEP carboxylation resulting in increased TCA cycle intermediates and a lower level of PEP was again clearly seen, but the level of pyruvate kinase was not significantly stimulated in this case, or if it was, then the utilization of pyruvate increased by the like amount so that no net increase in pyruvate was observed.

The fact that both in *Chlorella* and in isolated cells from poppy leaves addition of ammonium ion to cells respiring in the dark sometimes results in a rapid increase in glutamate labeling (in some cases there is an initial decrease followed by an increase) might seem to suggest that there must be a dark reaction for glutamate synthesis separate from the GOGAT reaction in the chloroplast which requires light. Probably, this labeling is due to increased availability of four- and five-carbon ketoacid skeletons labeled with $^{14}CO_2$ due to PEP carboxylation and transamination in which the amino group comes from pools of unlabeled amino acids and protein degradation.

Recently we have been able to isolate mesophyll cells from the leaves of Spinacea oleracea in such a way to allow high photosynthetic rates to be obtained even after the isolated cells have been maintained for as long as 60 hours. The incorporation of $^{14}CO_2$ into carbohydrates, carboxylic acids and amino acids in these cells and the influence of ammonia on that incorporation have been studied. The total incorporation, the specific radioactivity, and the pool sizes of most of most of the protein amino acids and gamma amino butyric acid have been determined as a function of time. By measuring specific activities and saturation effects, it has been possible to determine the presence and relative sizes of the metabolically active and passive pools for the various amino acids (Larsen et al. to be published). In general, the effects of added ammonium are similar to those seen earlier with Chlorella pyrenoidosa, with spinach leaf discs, with poppy cells, and with isolated spinach cells as reported recently by Woo and Canvin (27, 28). Reduced carbon from photosynthetic fixation is diverted away from sucrose synthesis and into the synthesis of amino acids. The labeling of malate, citrate, and in this case, of glycerate are considerably enlarged (Table 5).

The presence of ammonia during photosynthesis by these isolated spinach cells has profound effects on the pool sizes and specific activity of various amino acids (Table 6). However, detailed analysis of ^{14}C incorporation into individual amino acids showed that the ammonia effect is highly specific. Analysis of these effects in the light of present understanding of pathways of amino acid biosynthesis provides a better insight into the mode of action of ammonia, and the flow of nitrogen into the various amino acids.

Glutamine labeling increases 21-fold in cells treated with ammonia, while glutamate labeling decreases 30%, and at the same time there is a 5-fold increase in glutamine pool size and a 40% reduction in glutamate pool size (Table 6). The 4-fold increase in specific activity of glutamine is nearly matched by a similar increase in that of asparagine, but the pool size of asparagine remains constant while that of glutamine has increased 5-fold. Particularly dramatic is the 16-fold increase of ^{14}C incorporation into histidine accompanied by an increase of 10-fold in specific activity but only a modest change in pool size.

Table 5. Effect of NH_4^+ on incorporation of ^{14}C into carbohydrates and carboxylic acids in isolated spinach cells in the light.

Compound	Percent of incorporation obtained in cells with NO_3^- as N-source
Sucrose	70
Maltose	100
Glucose	90
Fructose	70
Malate	150
Citrate	190
Glycerate	140
Hexosemonophosphates	90
Phosphoglycerate	90
Phosphoenolpyruvate	110

Isolated mesophyll spinach cells were allowed to photosynthesize for 60 min (1850 $\mu E \cdot m^{-2} \cdot sec^{-1}$) at 20 C in 7 mM $H^{14}CO_3^-$.

In contrast, added ammonia has no effect on incorporation of ^{14}C into glycine, serine, phenylalanine and tyrosine. The lack of effect on glycine and serine is understandable in terms of their origin in the peroxisomes, where carbon ketoacid skeletons are dependent on glycolate formation or glycerate formation in the chloroplasts. These carbon skeleton origins bypass the reactions in the cytoplasm that are stimulated by ammonia addition. Phenylalanine and tyrosine may be less affected because of their dependence on the shikimic acid pathway located in the chloroplasts. Valine appears also to be formed in chloroplasts, as discussed in the first section, but for this amino acid, there is a strong stimulation in rate of labeling with added ammonia.

Table 6. Effect of NH_4^+ on biosynthesis of amino acids in spinach cells in the light

	% of value obtained with NO_3^- as N-source		
Amino Acid	Total Incorporation of ^{14}C	Specific Activity	Pool Size
Glutamate	70	120	60
Glutamine	2100	430	500
Aspartate	230	170	140
Asparagine	540	420	100
Alanine	150	120	120
Valine	330	310	110
Arginine	270	200	130
Histidine	1600	1000	160
Leucine	170	--	--
Glycine	90	100	90
Serine	90	90	100
Phenylalanine	80	80	110
Tyrosine	100	100	100
Threonine	100	230	50
Isoleucine	120	220	60
Proline	--	50	--
γ-Aminobutyrate	--	50	--

Conditions were as described for Table 5. Amino acids were separated by cation exchange chromatography followed by two-dimensional paper chromatography, radioautography and scintillation counting to determine ^{14}C incorporation. Portions of the samples were reacted with tritiated dansyl chloride after ion exchange chromatography, and the resulting derivatives were separated by polyamide thin-layer chromatography. Tritium/^{14}C ratios were then determined to obtain specific ^{14}C radioactivity after which pool sizes were calculated from total ^{14}C content and specific radioactivity. Detailed procedure to be published.

With all this evidence for a general shift in metabolism caused by external addition of ammonia, the next problem is to determine the mechanisms. This shift results in relatively greater synthesis of amino acids and less carbohydrates, especially sucrose. When there is a stimulation in CO_2 fixation, the extra fixed carbon is used in forming carbon skeletons for amino acid synthesis (28,29). In other cases, sucrose synthesis stops, is diminished, or sucrose is metabolized to provide carbon for amino acid synthesis (19,20,21,22,25). The shift seems to be somewhat independent of photosynthesis *per se*, since evidence is seen in the dark for stimulation of pyruvate kinase (and by inference PEP carboxylase)(20,22) with respiring *Chlorella pyrenoidosa* and for PEP carboxylase with respiring poppy cells (26).

The mechanism whereby ammonia added to the medium could cause these changes is difficult to determine. Given the low K_m for ammonium ion of glutamine synthetase, the level of free ammonia in cells would be expected to be kept very low. Nevertheless, the observation of a simultaneous shift in several rate-limiting steps linking photosynthetic carbon metabolism and amino acid synthesis seems meaningful. Perhaps ammonium ion in the medium induces internal changes that mimic an important regulatory process.

Externally applied ammonia might act in some indirect way. It could, for example, cause a change in the pH of the cytoplasm and this in turn could alter the activity of PEP carboxylase (26). In fact, Davies (29) proposed that PEP carboxylase acts as a cytoplasmic pH stat, increasing in activity with increasing pH, thus producing acid via carboxylation. If the signal (rise in pH) were caused by ammonia, the PEP carboxylation would supply oxaloacetate and oxoglutarate for glutamate synthesis and consequently for glutamine synthesis, using up the excess ammonia. The metabolic branch point in the cytoplasm for allocation of carbon to sucrose synthesis or to PEP formation *in the light* seems likely to be at triose phosphate coming from the chloroplast. The export of triose phosphate from the chloroplast is controlled by the level of P_i in the cytoplasm (30,31). The conversion of triose phosphate to hexose monophosphates may be controlled by the activity of cytoplasmic fructose 1,6-bisphosphatase. Subsequent sucrose formation could be regulated at the sucrose phosphate synthetase step.

Synthesis of glutamine upon ammonia addition uses up ATP, and could cause a drop in cytoplasmic ATP or energy charge. A decreased ATP level of ATP/ADP ratio could increase the conversion of triose phosphate to PGA and the conversion of PEP to pyruvate. Decreased ATP level could also lead to decreased UTP level.

The reaction of glucose-1-phosphate with UTP to make UDP-glucose (UDPG) is reversible (32), so this reaction would not itself be expected to become rate-limiting for sucrose synthesis. Nevertheless, a lower level of UDPG could decrease the activity of the enzyme for the following rate-limiting step, sucrose phosphate synthetase, which is reported to exhibit activity with a sigmoidal dependence on UDPG concentration (33). Also, it should be noted that sucrose phosphate synthetase is reportedly inhibited by sucrose (34). Isolated photosynthesizing poppy cells were found to accumulate sucrose due to the plasmodesmata being sealed on cell isolation (35). Thus, in our more recent experiments with isolated spinach cells, the four hours photosynthesis preceding incubation with $^{14}CO_2$ could have produced enough sucrose to inhibit sucrose phosphatase, and indirect effects of added ammonia could add to this inhibition. With freshly isolated spinach cells and high fixation rates (28,29), this inhibition of sucrose synthesis by ammonia addition might not be seen.

Increased conversion of triose phosphate to PGA, and subsequent reactions would of course produce ATP and NADH for biosynthetic reactions which might be stimulated by the increased supply of carbon and nitrogen metabolites. Thus one can envisage a general metabolic shift triggered by the increased supply of ammonium ions in the cytoplasm. Whether or not this is the actual or major mechanism for the effects of ammonia is a matter of conjecture at this time.

A substantial fraction of the total phosphate in the cytoplasm can be bound in sugar phosphates, UDPG, and even PP_i and any reduction in the levels of these metabolites in the cytoplasm accompanying decreased sucrose synthesis could result in an increased level of P_i in the cytoplasm. This increased P_i level in turn would stimulate triose phosphate export from the chloroplast.

The P_i entering the chloroplast would inhibit starch formation (36) and could stimulate CO_2 fixation if the previous level of P_i had been suboptimal for maximum photophosphorylation. This would account for the sometimes observed increase in CO_2 fixation seen on addition of ammonia.

SUMMARY

Leaf cell chloroplasts have the competence to make most, if not all, amino acids required for protein synthesis. Aside from amino acids involved in the photorespiratory nitrogen cycle, chloroplasts play a quantitatively important role in the synthesis of amino acids derived from the shikimic acid pathway and of valine. Synthesis of other amino acids by isolated chloroplasts appears to be relatively minor compared to the cells synthesis as a whole, but this could be due to lack of optimal physiological conditions for the isolated organelles.

Ketoacid skeletons for serine and glycine synthesis are being made in peroxisomes. Skeletons for glutamate and glutamine are found in mitochondria, and aspartate and asparagine skeletons are made in the cytoplasm. These ketoacids need not be made, therefore, in chloroplasts. It now appears that chloroplasts can make most other required ketoacid skeletons, whether of not they are made elsewhere as well.

The glycolate pathway is bidirectional with respect to serine and glycine formation but net carbon flow in the direction glycerate to serine to glycine seems to occur (at least in isolated spinach cells) only with high CO_2 (0.2%)

and low O_2 (2%). With inhibition of the glycine to serine conversion by glycine hydroxamate, diminished ammonia evolution results in accumulation of amino groups in glycine and serine and a decrease in glutamate and other amino acids.

Ammonia added to a variety of photosynthetic cells in liquid suspension shifts biosynthetic metabolism away from carbohydrate (especially sucrose) synthesis and towards synthesis of certain (but not all) amino acids. A fuller understanding of this shift may shed light on the way plant cells control the flow of photosynthate to biosynthesis in response to physiological conditions.

ACKNOWLEDGEMENT

This work was supported in part by the Division of Biological Energy Conversion and Conservation, Office of Basic Energy Sciences, Department of Energy, under Contract No. W-7405-ENG-48, and in part by a Rockefeller Foundation grant to A.L.L. and a Danish Research Council grant to P.O.L.

REFERENCES

1. Lea, PJ, and BJ Miflin, Photosynthetic ammonia assimilation, In: Encyclopedia of Plant Physiology, New Series Volume 6 Photosynthesis II, Gibbs, M and E Latzko (eds.), Berlin, Springer, 445-456, 1979.
2. Lea, PJ and BJ Miflin, Alternative route for nitrogen assimilation in higher plants. Nature (London) 251: 614-616, 1974.
3. Wallsgrove, RM, PJ Lea, and BJ Miflin, Distribution of the enzymes of nitrogen assimilation within the pea leaf. Plant Physiol 63:232-236, 1979.
4. Keys, AF, IF Bird, MJ Cornelius, PJ Lea, RM Wallsgrove, and BJ Miflin, Photorespiratory nitrogen cycle. Nature 275:741-743, 1978.
5. Mitchell, CA and CR Stocking. Kinetics and energetics of light-driven chloroplast glutamine synthesis. Plant Physiol 55:59-63, 1975.
6. Givan, CV and RM Leech. Biochemical autonomy of higher plant chloroplasts and their synthesis of small molecules. Biol Rev 46:409-428, 1971.
7. Jensen, RG and JA Bassham, Photosynthesis by isolated chloroplasts. Proc. Natl Acad Sci (US) 56:1095-1101, 1966.

8. Kirk, PR and RM Leech, Amino acid biosynthesis by isolated chloroplasts during photosynthesis. Plant Physiol 50:228-234, 1972.

9. Buchholz, B, B Reupke, H. Bickel, and G Schultz, Reconstitution of amino acid synthesis by combining spinach chloroplasts with other leaf organelles. Phytochemistry 18:1109-1111, 1979.

10. Larsson, C and PA Albertson, Photosynthetic $^{14}CO_2$ fixation by chloroplast populations isolated by a polymer two-phase technique. Biochim Biophys Acta 357:412-419, 1974.

11. Bickel, H and G Schultz, Shikimic pathway regulation in suspensions of intact spinach chloroplasts. Phytochemistry 18:498-499, 1979.

12. Platt, SG, Z Plaut, and JA Bassham, Analysis of steady state photosynthesis in alfalfa leaves. Plant Physiol 57:69-73, 1976.

13. Frederick, SE and EH Newcomb, Cytochemical localization of catalase in leaf microbodies (peroxisomes). J Cell Biol 43:343-353, 1969.

14. Rabson, R., NE Tolbert, PC Kearney, Formation of serine and glyceric acid by the glycolate pathway. Arch Biochem Biophys 98:154-163, 1962.

15. Tolbert, NE, Microbodies-peroxisomes and glyoxysomes. Annu Rev Plant Physiol 22:45-74, 1971.

16. Randall, DD, NE Tolbert, and D Gremel, 3-phosphoglycerate phosphatase in plants. II Distribution, physiological considerations and comparison with P-glycolate phosphatase. Plant Physiol 48:480-487, 1971.

17. Platt, SG, Z Plaut, and JA Bassham. Steady state photosynthesis in alfalfa leaflets. Effects of CO_2 concentration. Plant Physiol 60:230-234, 1977.

18. Lawyer, AL and I Zelitch, Inhibition of glycine decarboxylation and serine formation in tobacco by glycine hydroxamate and its effect on photorespiratory carbon flow. Plant Physiol 64:706-711, 1979.

19. Kanazawa, T, K Kanazawa, MR Kirk and JA Bassham, Regulation of photosynthetic carbon metabolism in synchronously growing Chlorella pyrenoidosa. Plant Cell Physiol 11:149-160, 1970.

20. Kanazawa, T, K Kanazawa, MR Kirk, and JA Bassham, Difference in nitrate reduction in "light" and "dark" stages of synchronously grown Chlorella pyrenoidosa and resultant metabolic changes. Plant Cell Physiol 11:445-452, 1970.

21. Kanazawa, T, MR Kirk and JA Bassham, Regulatory effects of ammonia on carbon metabolism in photosynthesizing Chlorella pyrenoidosa. Biochim Biophys Acta 205:401-408, 1970.

22. Kanazawa, T., K Kanazawa, MR Kirk and JA Bassham, Regulatory effects of ammonia on carbon metabolism in Chlorella pyrenoidosa during photosynthesis and respiration. Biochim Biophys Acta 226:656-669, 1972.

23. Platt, SG, Z Plaut, and JA Bassham, Ammonia regulation of carbon metabolism in photosynthesizing leaf discs. Plant Physiol 60:739-742, 1977.
24. Paul, JS and JA Bassham, Maintenance of high photosynthetic rates in mesophyll cells isolated from Papaver somniferum. Plant Physiol 60:775-778, 1977.
25. Paul, JS, KL Cornwell, and JA Bassham, Effects of ammonia on carbon metabolism in photosynthesizing isolated mesophyll cells from Papaver somniferum L. Planta 142:49-54, 1978.
26. Hammel, KE, KL Cornwell, and JA Bassham, Stimulation of dark CO_2 fixation by ammonia in isolated mesophyll cells of Papaver somniferum L. Plant Cell Physiol 20:1523-1529, 1979.
27. Woo, KC and DT Canvin, Effects of ammonia on photosynthetic carbon fixation in isolated spinach leaf cells. Canadian J Bot 58:505-510, 1980.
28. Woo, KC and DT Canvin, Effects of ammonia, nitrate, glutamate, and inhibitors of nitrogen metabolism on photosynthesis in isolated spinach leaf cells. Canadian J Bot 58:511-516, 1980.
29. Davies, DD, Control of and by pH, In: Rate Control of Biological Processes, Davies, DD (ed.) Cambridge, 513-529, 1973.
30. Heldt, HW and L Rappley, Specific transport of inorganic phosphate, 3-phosphoglycerate and dihydroxyacetonephosphate, and of dicarboxylates across the inner membrane of spinach chloroplasts. FEBS Lett 10:143-148, 1970.
31. Fliege, R, UI Flugge, K Werdan, and HW Heldt, Specific transport of inorganic phosphate, 3-phosphoglycerate and triose-phosphate across the inner membrane of the envelope in spinach chloroplasts. Biochim Biophys Acta 502:232-247, 1978.
32. Gustafson, GL and JE Bander, Uridine diphosphate glucose pyrophosphorylase from Sorghum vulgare. Purification and kinetic properties. J Biol Chem 247:1387-1397, 1972.
33. Preiss, J and E Greenberg, Allosteric regulation of uridine diphosphoglucose: D-fructose-6-phosphate-2-glucosyl transferase (E.C.2.4.1.14) Biochem Biophys Res Commun 36:289-295, 1969.
34. Salerno, GL and HG Pontis, Studies on sucrose phosphate synthetase. The inhibitory action of sucrose. FEBS Lett 86:263-267, 1978.
35. Kaiser, WM, JS Paul and JA Bassham. Release of photosynthates from mesophyll cells in vitro and in vivo. Z Pflansenphysiol 94:377-385, 1977.
36. Ghosh, HP and J Preiss, Adenosine diphosphate glucose pyrophosphorylase. A regulatory enzyme in the biosynthesis of starch in spinach leaf chloroplasts. J Biol Chem 291:4491-4505, 1966.

CHAPTER 6

PHOTOSYNTHATE LIMITATION OF SYMBIOTIC N_2 FIXATION

K. Andersen, S.T. Lim, H. Spiller, K.T. Shanmugam, and R.C. Valentine*

*Plant Growth Laboratory, Agronomy and Range Science Dept., University of California at Davis, CA 95616, U.S.A.

Introduction

A photorespiration-like system in autotrophic bacteria as a model for determining the *in vivo* significance of the oxygenase function of ribulose biphosphate carboxylase in photosynthate production

Efficiency of photosynthate utilization coupled to symbiotic N_2 fixation in legumes

Environmental factors

Is nitrogenase-mediated H_2 evolution a constant or variable function of the enzyme?

Do cyclic mononucleotides (cAMP and cGMP) function as chemical messengers in rhizobial bacteroids integrating availability and efficiency of utilization of photosynthate?

Photosynthate limitation in free-living, heterocystous blue-green algae as a model for symbiotic N_2 fixation by the *Azolla-Anabaena* symbiosis

References

J.D. Bewley (ed.), Nitrogen and Carbon Metabolism

INTRODUCTION

For the past several years our laboratory has studied various aspects of the energetics of nitrogen fixation, experiments leading to the conclusion that there is a high metabolic energy cost for this process (1). Several of these experiments will be discussed below (see Figs. 1 and 4 for integration of this work). Subsequently, research focused on enhancing the energy efficiency of symbiotic N_2 fixation has been emphasized. Important questions are:

1. How significant is the oxygenase function of the host plant ribulose bisphosphate carboxylase in terms of efficiency of photosynthate production supporting symbiotic N_2 fixation? Can the oxygenase function be altered by mutations?
2. What parameters control the efficiency of photosynthate conversion linked to symbiotic nitrogen fixation? Is the nitrogenase-mediated H_2 evolution of rhizobial bacteroids a constant or a variable function?
3. Do cyclic mononucleotides (cAMP/cGMP) function as chemical messengers integrating availability and efficiency of utilization of photosynthate in nodules?
4. Is photosynthate a limiting factor for nitrogen fixation in heterocystous blue-green algae and their symbiotic associations (e.g., *Azolla*)?

A PHOTORESPIRATION-LIKE SYSTEM IN AUTOTROPHIC BACTERIA AS A MODEL FOR DETERMINING THE *IN VIVO* SIGNIFICANCE OF THE OXYGENASE FUNCTION OF RIBULOSE BISPHOSPHATE CARBOXYLASE IN PHOTOSYNTHATE PRODUCTION

As illustrated in Fig. 1, the host plant must supply all of the energy as photosynthetically fixed carbon (photosynthate) for supporting the high metabolic energy demand of symbiotic N_2 fixation. There is increasing evidence that available photosynthate may limit symbiotic N_2 fixation in leguminous crops such as soybean (2). This notion was dramatized by

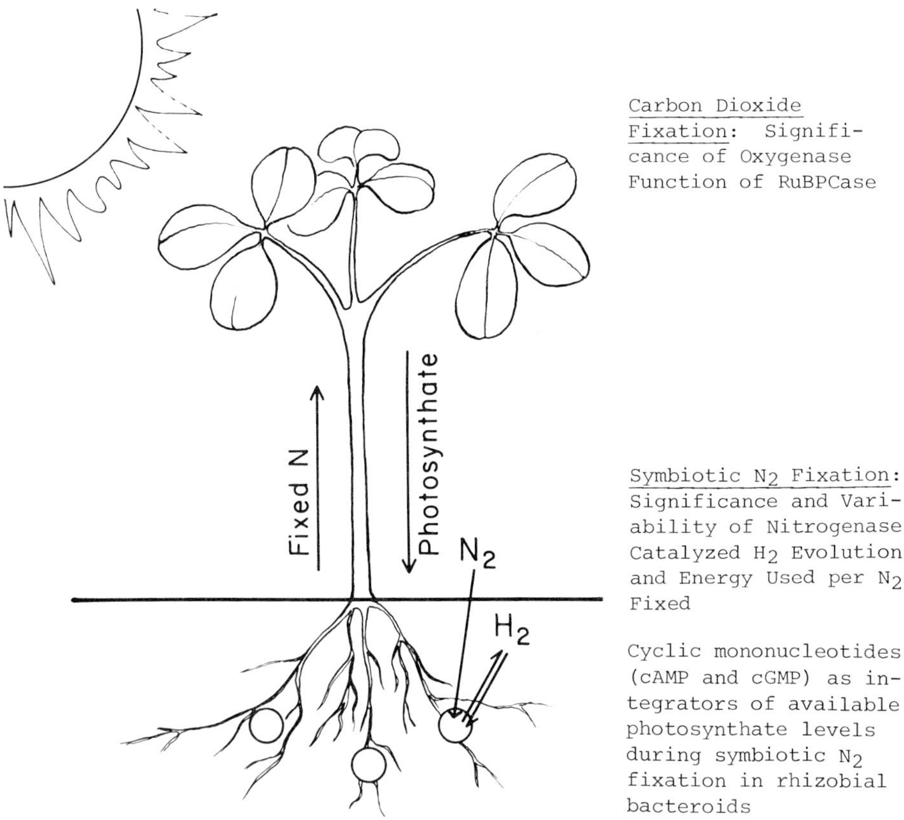

Figure 1: Symbiotic nitrogen fixation by a leguminous plant. Fixed carbon (photosynthate) derived from photosynthesis in the leaf is translocated to the nodule where it powers nitrogen fixation in the Rhizobium symbiont, which, in turn, feeds the plant fixed nitrogen.

the "CO_2 fertilization" experiments of Hardy and Havelka (3). Since it is not yet economically feasible to directly exploit the concept of enhanced N_2 fixation by CO_2 fertilization of crop lands, an alternative approach is to genetically manipulate the limiting step(s) of photosynthate production. As part of our research program on the energy supply linked to N_2 fixation we have investigated the efficiency of CO_2 fixation with an eye toward the significance and possible genetic alteration of the oxygenase function of ribulose bisphosphate (RuBP) carboxylase.

The experimental tool chosen for these studies is a simple bacterium, *Alcaligenes eutrophus*, possessing a plant-type RuBP carboxylase/oxygenase. This bacterium, which induces an active glycollate oxidizing pathway during CO_2 fixation, was chemically mutated leading to a complete blockage of glycollate catabolism. Mutant strain AE161, which lacked glycollate oxidoreductase activity, excreted up to 1.2 µmol glycollate/mg cell protein per h during autotrophic growth. This mutant strain was used to study the efficiency of CO_2 fixation in terms of how much of the fixed carbon was excreted as glycollate under different conditions (4). Glycollate excretion was not detected during heterotrophic growth. Only 1% of the total CO_2 fixed was excreted as glycollate in an atmosphere of 4% CO_2 plus 20% O_2 (Fig. 2). The rate of glycollate excretion showed a large increase, and CO_2 fixation decreased as the CO_2 concentration was lowered. Almost half (45-50%) of the total CO_2 fixed was excreted as glycollate in an atmosphere of 0.07% CO_2 plus 20% O_2.

Increased glycollate excretion was also observed when the O_2 concentration was increased in the presence of a constant CO_2 concentration.

We conclude from these studies that the ratio of CO_2/O_2 surrounding the cell plays a profound role in determining the efficiency of CO_2 fixation in *A. eutrophus*, analogous to the situation for C_3 plants. These effects are believed to be mainly due to the oxygenase activity of RuBP carboxylase. The efficiency of CO_2 fixation might be increased by (a) increasing the ratio of CO_2/O_2 at the catalytic site

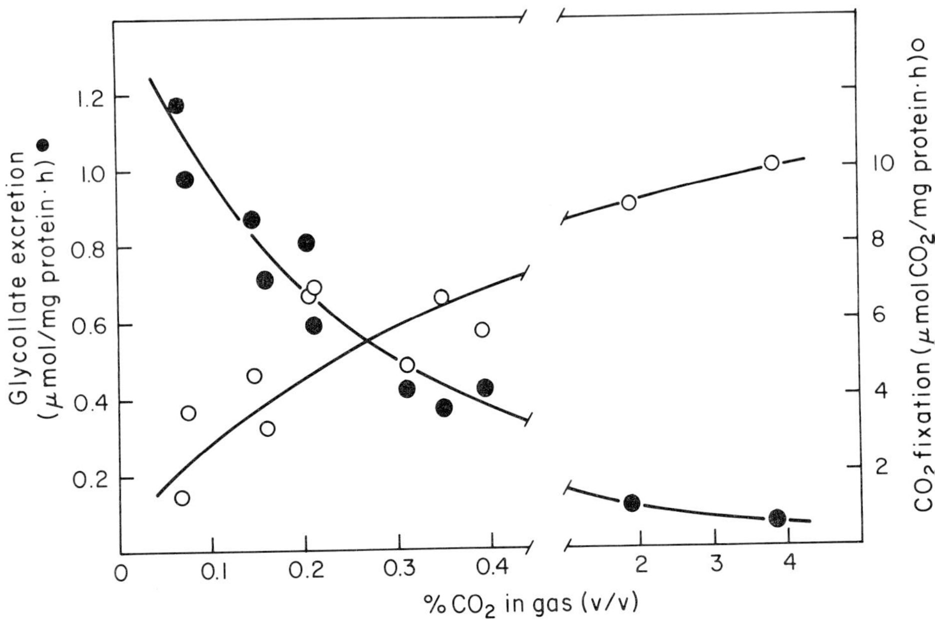

Figure 2: Net CO_2 fixation and glycollate excretion by A. eutrophus mutant strain AE161 as a function of CO_2 concentration in the presence of 20% O_2 (King and Andersen, Reference 4).

of RuBP carboxylase/oxygenase (as is apparently done by the C_4 pathway in C_4 plants, or by other CO_2 concentrating systems such as may occur in microorganisms; or (b) reducing the oxygenase activity of RuBP carboxylase. Thus there is considerable incentive for studies aimed at genetic enhancement of photosynthetic CO_2 fixation, work now underway with the isolation and characterization of the first set of catalytic mutants of RuBP carboxylase/oxygenase in *A. eutrophus* (5). Recent work has been focused on the development of genetic systems to allow fine structure mapping of RuBP carboxylase genes and their mutants in *A. eutrophus* (6) -- experiments aimed at pinpointing important catalytic and regulatory functions of this important enzyme.

Environmental Factors

The efficiency of utilization of photosynthate by the bacteroids in the nodulĕ is of great interest, and many studies are aimed at maximizing the amount of N_2 fixed per unit photosynthate consumed to conserve available energy in the nodule.

What parameters control the efficiency of photosynthate conversion linked to N_2 fixation *in vivo*? Studies with free-living N_2-fixing bacteria may partly answer this question. We have used nitrogenase-derepressed mutant strains of *Klebsiella pneumoniae* to study the energetics of N_2 fixation (7). These studies showed that the efficiency of N_2 fixation in living cells measured in terms of moles of carbohydrate consumed per mole of N_2 fixed, or in terms of

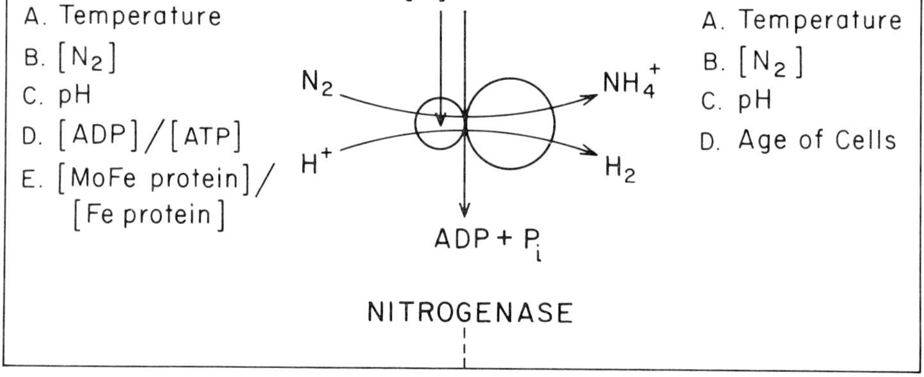

Figure 3: Factors affecting the efficiency of N_2 fixation in terms of energy consumed in the nitrogenase reaction per N_2 fixed. Efficiency may be decreased by increasing the amount of ATP hydrolyzed per electron transferred and/or by increasing the amount of H_2 formed per N_2 fixed. For in vitro experiments see recent reviews (12, 13) for references.

Table 1. Influence of environmental conditions on the efficiency of N_2 fixation in the nitrogenase-derpressed mutant N-20 of *Klebsiella pneumoniae*. See Andersen and Shanmugam (1977) for experimental details.

Temperature (°C)	pH	Age of Culture (days)	Glucose Consumed* / NH_4^+ Produced	Apparent ATP/N_2 Value	Nitrogenase Catalyzed H_2 Evolution* $\frac{H_2}{NH_4^+}$
25	7.3	2	3.9	23	0.63
40	7.3	2	6.3	30	1.1
25	4.8	2	8.0	41	2.0
25	7.3	6	7.9	47	0.65

*molar ratio

ATP/N_2, is very sensitive toward a number of environmental and physiological parameters (reference 7, Fig. 3 and Table 1). From these studies it is interesting to speculate that the efficiency of photosynthate consumption coupled to symbiotic N_2 fixation in nodules will vary markedly depending on temperature, osmotic strength, pH, age of nodules, etc. For example, salt has a marked influence on N_2-fixation in *Klebsiella in vivo*, and the relationship between N_2 fixation and osmoregulation is discussed in a recent paper (8).

Is Nitrogenase-Mediated H_2 Evolution a Constant or Variable Function of the Enzyme?

A significant proportion ($\geq$ 1/3) of the energy consumed during N_2 fixation *in vivo* is lost as H_2 evolution mediated by nitrogenase (7). If this serious side-reaction is to be optimally managed during crop production it is important to know what conditions significantly influence H_2 production. This refers to H_2 production catalyzed by the hydrogenase function of nitrogenase; H_2 utilization or uptake, a separate reaction, is discussed elsewhere (9). It is interesting to compare the H_2 forming function of nitrogenase with the oxygenase function of carboxylase, discussed above. In the case of carboxylase more efficient CO_2 fixation can be achieved *in vivo* by simply manipulating the ratio of CO_2/O_2 over the cells, and consequently, as seen by the enzyme. The two competing substrates for nitrogenase are protons (H^+) and N_2. Sparging of N_2-fixing cells with an inert gas to remove all traces of N_2 leads to greatly accelerated H_2 evolution. However, several atmospheres of N_2 over N_2-fixing cells does not significantly diminish H_2 evolution below the level at 1 atm N_2. On the other hand, because of the relatively high affinity of nitrogenase for its substrate N_2 and the ubiquitous nature of N_2 gas, it seems unlikely that N_2 limitation is a significant parameter controlling H_2 evolution. However, studies summarized in Table 1 and Fig. 3 show that other environmental parameters such as increased temperature or low pH may adversely affect H_2 evolution *in vivo*. This aspect deserves further work with the future goal of minimizing H_2 evolution catalyzed by nitrogenase. Some of the energy lost through H_2 evolution may be recovered by recycling through uptake hydrogenases. Unfortunately, many agronomically important *Rhizobium* strains lack H_2 uptake (Hup) activity. We have recently discussed the possibilities for genetic engineering of Hup^+ *Rhizobium* strains (9).

DO CYCLIC MONONUCLEOTIDES (cAMP AND cGMP) FUNCTION AS CHEMICAL MESSENGERS IN RHIZOBIAL BACTEROIDS INTEGRATING AVAILABILITY AND EFFICIENCY OF UTILIZATION OF PHOTOSYNTHATE?

Recently it has been found that utilization of H_2 by *Rhizobium* strains is dependent on the nature of the carbon source present in the medium. When exposed to either fructose or glucose plus H_2, *R. japonicum* grew at the expense of both energy sources, while compounds like malate, succinate or fumarate inhibited H_2 utilization. The addition of 1 mM cAMP overcame the inhibition by malate of the utilization of H_2 by a mechanism that appeared to require both RNA and protein synthesis, as evidenced by the sensitivity to the antibiotics rifampicin and chloramphenicol. Measurements in malate-grown *R. japonicum* of the intracellular concentration of cAMP by radioimmunoassay indicated a low concentration of cAMP. When glutamate was present as the sole carbon and nitrogen source, the high uptake activity of glutamate and high Hup activity coincided with a high intracellular concentration of cAMP (10). The results suggest that in the presence of cAMP, proteins can be synthesized which allow the metabolism of other energy sources, such as H_2. So malate (and probably other organic acids) may serve as catabolite repressors of H_2 utilization in *R. japonicum* in a manner analogous to the repression by glucose of lactose utilization in *Escherichia coli* (11).

cGMP considerably reduced hydrogenase activity when added at concentrations as low as 0.1 mM to cultures of *R. japonicum*. The overall rate of protein synthesis and the activity of other selected enzymes, such as glutamine synthetase and malate dehydrogenase, were not significantly influenced (12). Experiments with *R. japonicum* have shown that intracellular concentrations of cGMP varied 10-fold during growth under different O_2 concentrations. The amount of cGMP may be under redox control, such that a decrease in oxygen concentration lowers the cGMP concentration, allowing the expression of *hup* and genes for other enzyme systems active under nitrogen-fixing conditions (12).

Based on these experiments it is interesting to speculate that both the quality and quantity of photosynthate supplied by the host plant to their rhizobial bacteroids may affect the function of H_2 uptake as well as other enzyme systems involved in efficient utilization of photosynthetically derived energy sources. There may exist a dynamic control mechanism mediated by cyclic mononucleotides which integrates photosynthate supply in the nodules and demand for symbiotic N_2 fixation.

PHOTOSYNTHATE LIMITATION IN FREE-LIVING, HETEROCYSTOUS BLUE-GREEN ALGAE AS A MODEL FOR SYMBIOTIC N_2 FIXATION BY THE *AZOLLA-ANABAENA* SYMBIOSIS

The *Azolla-Anabaena azollae* symbiosis represents a unique "leaf nodule" design for efficient photosynthetic N_2 fixation, a system utilized in Asia on a large scale for N_2 fixation in paddy rice culture. In order to achieve such high levels of N_2 fixation the symbiotic blue-green algae must be provided with an optimum environment by the host fern plant. In the experiments described below attention is focused on the role of exogenously supplied carbon sources ("photosynthate") for achieving a stimulation of N_2 fixation by free-living blue-green algae.

The heterocystous filamentous blue-green alga *Anabaena variabilis* is capable of growing heterotrophically in the dark at the expense of added fructose. Since it can also grow autotrophically in light deriving its carbon from CO_2 fixation, it presents an integrated model of symbiosis with the heterocyst manufacturing fixed nitrogen and the vegetative cell taking up fructose or supplying fixed

Table 2. Stimulation by fructose or CO_2 enrichment of nitrogenase activity and growth rates in *Anabaena variabilis*.

Previous Growth Medium of Inoculum	Present Growth Medium	Doubling Time (h)	Nitrogenase μmol C_2H_4/h x OD_{750}
air	air	22	.29
air	air/.4% CO_2	17.6	.45
air	air/fructose	18.7	.51
fructose/air	air	15.3	.41
fructose/air	air/fructose	12	.74

Cultures were grown under continuous light (3000 Lux) in 125 ml Erlenmeyer flasks. Inoculum was $OD_{750} \leq 0.03$ from mature cultures of either air or fructose (10 mM) grown cells. Samples for determination of doubling time and nitrogenase activity were taken from days 2-4 after inoculum, and acetylene reduction determined for 1.5 h. OD_{750} was determined after sonication and proper dilution of the algal suspension.

carbon for the heterocyst. Under the growth conditions described in Table 2 growth was nitrogen-limited with a doubling time of 22 h in air with ambient CO_2 (0.03%). Supplementation with either CO_2 (0.4%) or fructose (10 mM) not only reduced doubling time to 17 h but also increased N_2-fixation by 72%. It appears that air-grown cultures immediately respond to increased CO_2 concentration or addition of fixed carbon by accelerating photosynthetic N_2-fixation. Growth rate could be even more enhanced if the culture previously had been grown in fructose/air. In this case, doubling time dropped to 12 h and nitrogenase activity peaked at 255%. The influence of the previous growth medium (fructose) enabling the vegetative cells to store higher glycogen reserves suggests that possibly both the supply of fixed carbon as well as reducing power may control the nitrogenase activity. Furthermore, much less O_2 will be evolved in the presence of fructose as compared

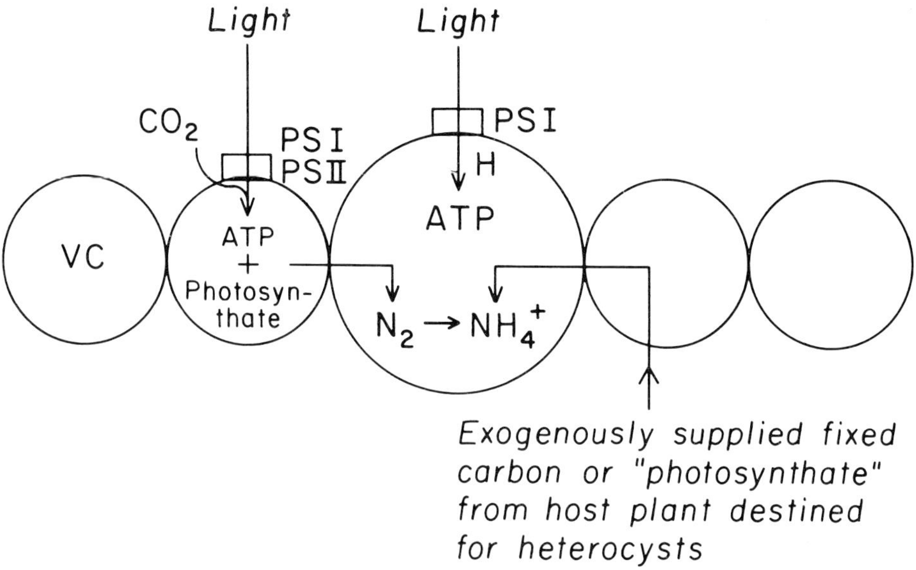

Figure 4: Model illustrating "photosynthate" limitation of N_2 fixation catalyzed by heterocystous blue-green algae. Maximum rates of N_2 fixation by free-living algae are achieved by supplying an exogenous source of fixed carbon such as sugars. It is hypothesized that in the case of the Azolla-Anabaena azollae association photosynthate supplied by the host fern enhances the rates of N_2 fixation by the symbiotic blue-green algal partner. VC, vegetative cells; H, heterocysts.

to CO_2 alone, thereby minimizing the oxygen inhibition of nitrogenase. It has been shown that ^{14}C from ^{14}C-labelled fructose rapidly accumulates in the heterocysts (data presented elsewhere). Thus, one may conclude that catabolism of a monosaccharide such as fructose which possibly could be a product from photosynthesis ("photosynthate") (a) influences O_2 levels in filaments, (b) provides carbon and (3) maintains a certain electron pressure in heterocysts (redox control), thereby regulating nitrogenase activity in blue-green algae.

REFERENCES

1. Shanmugam, KT, F O'Gara, K Andersen, and RC Valentine. Biological nitrogen fixation. Ann Rev Plant Physiol 29:263-276, 1978

2. Phillips, DA. Efficiency of symbiotic nitrogen fixation in legumes. Ann Rev Plant Physiol 31:29-49, 1980

3. Hardy, RWB and UD Havelka. Nitrogen fixation: a key to world food? Science 188:633-643, 1975

4. King, WR and K Andersen. Efficiency of CO_2 fixation in a glycollate oxidoreductase mutant of *Alcaligenes eutrophus* which exports fixed carbon as glycollate. Arch Microbiol in press, 1980

5. Andersen, K. Mutations altering the catalytic activity of a plant-type ribulose bisphosphate carboxylase/oxygenase in *Alcaligenes eutrophus*. Biochem Biophys Acta 585 1-11, 1979

6. Andersen, K. Biochemical genetics of ribulose bisphosphate carboxylase/oxygenase: high frequency transfer of and chromosome mobilization by broad host-range R-factors in CO_2 fixing bacteria, In: Genetic Engineering of Symbiotic Nitrogen Fixation and Conservation of Fixed Nitrogen, Rains, DW *et al.*, Plenum Press, in press, 1980

7. Andersen, K and KT Shanmugam. Energetics of biological nitrogen fixation: determination of the ratio of formation of H_2 to NH_4^+ catalyzed by nitrogenase of *Klebsiella pneumoniate in vivo*. J Gen Microbiol 103:107-122, 1977

8. Rudulier, DL, S Yang, and LN Csonka. Proline-overproduction enhances nitrogenase activity under osmotic stress in *Klebsiella pneumoniae*, In: Genetic Engineering of Symbiotic Nitrogen Fixation and Conservation of Fixed Nitrogen, Rains, DW *et al.*, Plenum Press, in press, 1980

9. Lim, ST, K Andersen, R Tait, and RC Valentine. Genetic engineering in agriculture: hydrogen uptake (*hup*) genes. Trends Biochem Sci 5:167-170, 1980

10. Lim, ST and KT Shanmugam. Regulation of hydrogen utilization in *Rhizobium japonicum* by cyclic AMP. Biochem Biophys Acta 584:479-492, 1979

11. Lewin, B (ed.). Gene expression, In: Bacterial Genomes, Vol 1, London, Wiley, 302-306, 1974

12. Lim, ST, H Hennecke and DB Scott. Effect of guanosine 3',5'-cyclic monophosphate on nitrogen fixation in *Rhizobium japonicum*. J Bacteriol 139:256-273, 1979

13. Winter, HC and RH Burris. Nitrogenase. Ann Rev Biochem 45:409-426, 1976

14. Mortenson, LE and RNF Thorneley. Structure and function of nitrogenase. Ann Rev Biochem 48:287-418, 1979

CHAPTER 7

PHOTORESPIRATION AND NITROGEN METABOLISM

David T. Canvin
Department of Biology, Queen's University,
Kingston, Ontario K7L 3N6

J.D. Bewley (ed.), Nitrogen and Carbon Metabolism

Photorespiration occurs in all C_3 higher plants as a result of the formation of phosphoglycolate and the subsequent metabolism of glycolate through the glycolate pathway (Fig. 1). It has as its external manifestations O_2 uptake and CO_2 release. The O_2 uptake observed in the light (3, 8) can be entirely (3) or largely (8) attributed to oxygenation of ribulose-1,5-bisphosphate and the subsequent metabolism of glycolate. The CO_2 production arises from the conversion of glycine to serine (7) or possibly some may arise from the decarboxylation of glyoxylate (10). Nitrogen metabolism is associated with photorespiration at several points: 1) the transamination of glyoxylate with glutamate to form glycine; 2) the release of ammonia during glycine decarboxylation; 3) the reassimilation of this released ammonia into amino acids; 4) the transamination of glyoxylate with serine to yield hydroxy-pyruvate and glycine.

The aminotransferases of reactions 1 and 4 (Fig. 1) have been studied by Rehfeld and Tolbert (20). Both a glutamate:glyoxylate aminotransferase and a serine:glyoxylate aminotransferase are present in the peroxisomes. Both enzymes are irreversible, i.e. they only catalyze the formation of glycine and will not transfer the amino group from glycine. The serine:glyoxylate aminotransferase has a Km for glyoxylate of 0.15 mM whereas the glutamate:glyoxylate aminotransferase has a Km (glyoxylate) of 4.4 mM. Since the respective Kms for serine and glutamate are similar, the amino group transfer from serine would be favoured when serine was present. The glutamate:glyoxylate aminotransferase must still operate, though, as two molecules of glycine must be produced for every serine that is formed. Based on the specific activities of the enzymes in isolated peroxisomes they are both present in sufficient

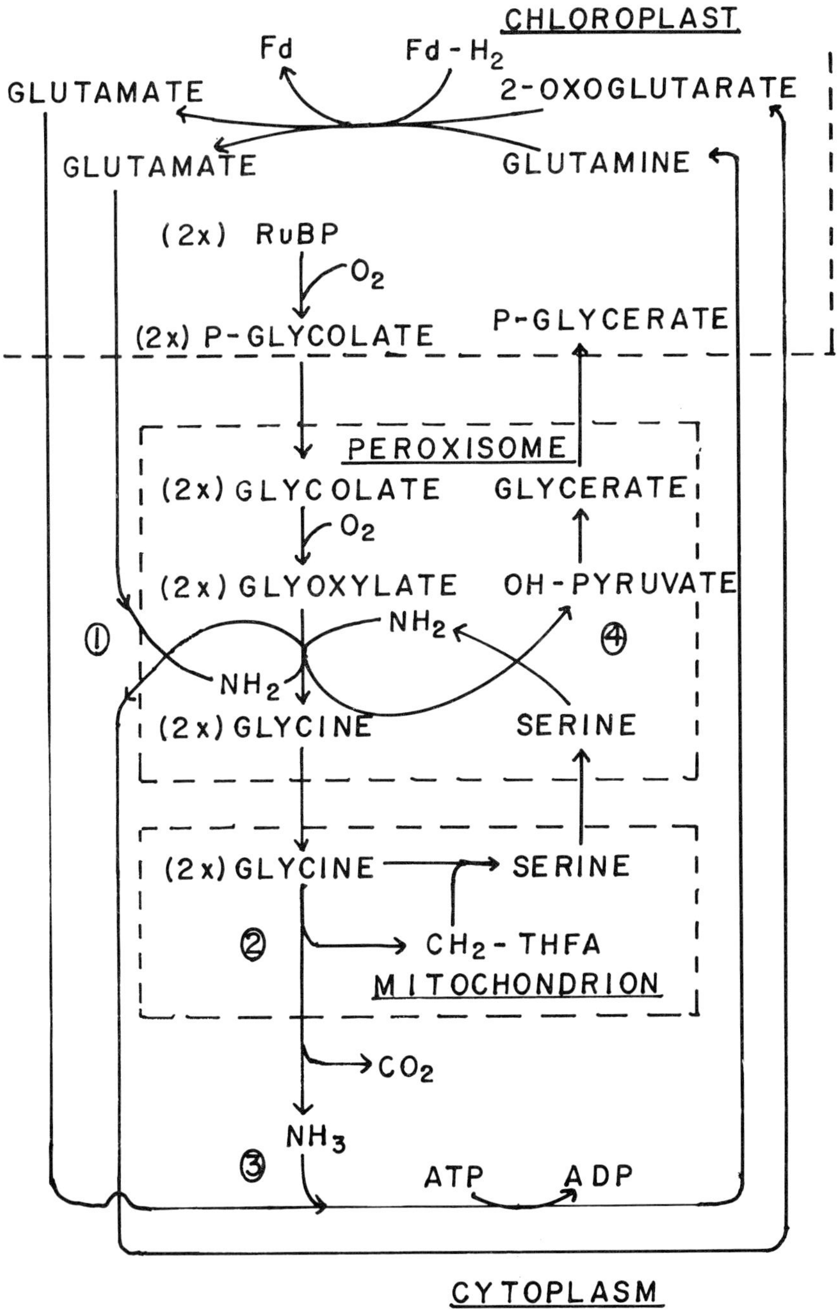

Figure 1. The photorespiratory nitrogen cycle. After Keys et al. (13).

quantities to cope with the flux of carbon through the glycolate pathway (24).

The operation of the glutamate:glyoxylate aminotransferase would require the continuous availability of glutamate in the peroxisome. It seems well-established that the principal site of 2-oxoglutarate use and glutamate formation in the leaf is the chloroplast (Fig. 1). The transfer of the amino nitrogen between chloroplast and peroxisome may involve glutamate and 2-oxoglutarate directly or the transfer may be mediated via an aspartate:oxaloacetic acid shuttle. Both chloroplasts and peroxisomes contain an aspartate:2-oxoglutarate aminotransferase (20) and the nitrogen may be transferred from the chloroplast to the peroxisome as aspartate with the carbon returning to the chloroplast as oxaloacetic acid. If the latter were the case, glutamate would be formed in the peroxisome by the aspartate:2-oxoglutarate aminotransferase and 2-oxoglutarate produced in the chloroplast by the enzyme in that organelle.

The essentiality of the serine:glyoxylate aminotransferase to the operation of the glycolate pathway is further shown by the studies of Somerville and Ogren (22) with mutants of *Arabidopsis* that are lacking in this enzyme. The mutants grow phenotypically normal under 1% CO_2 where the production of phosphoglycolate is inhibited. Under normal atmospheric conditions, however, photosynthesis was inhibited 79% after 30 min. After 10 min of $^{14}CO_2$ supply under normal atmospheric conditions the amount of ^{14}C in glycine was doubled and the amount in serine increased 2.5 fold over that in wild type plants. Compelling genetic, as well as biochemical, evidence supports the conclusion that the conditional lethal phenotypes of the mutants is solely due to the absence of serine-glyoxylate aminotransferase. The reason for the inhibition of photosynthesis and the lethal condition in air is not entirely clear but it has been suggested that the lack of reentry of photorespiratory carbon into the Calvin cycle may have a regulatory effect on photosynthesis and over long periods of time lead to a net carbon deficit (22).

Glycine decarboxylation (reaction 3, Fig. 1) has been investigated by several groups (4, 5, 14, 28, 29, 30). The enzyme is located on the inner surface of the inner membrane of the mitochondrion (29) and catalyzes the conversion of glycine to CO_2, NH_3 and a methylene-N^5, N^{10}-tetrahydrofolate unit (Fig. 2). The C_1 unit is utilized along with another molecule of glycine by serine hydroxymethyltransferase to produce serine (31). The net result of the combined action of these two enzymes is the conversion of 2 glycines to serine, CO_2, NH_3 and NADH (Fig. 2).

In isolated mitochondria where the energy charge is low the NADH may be oxidized through the electron transport chain to produce 3 ATP (4, 5) or it may also reduce oxaloacetate to malate in the mitochondrion (28). This malate

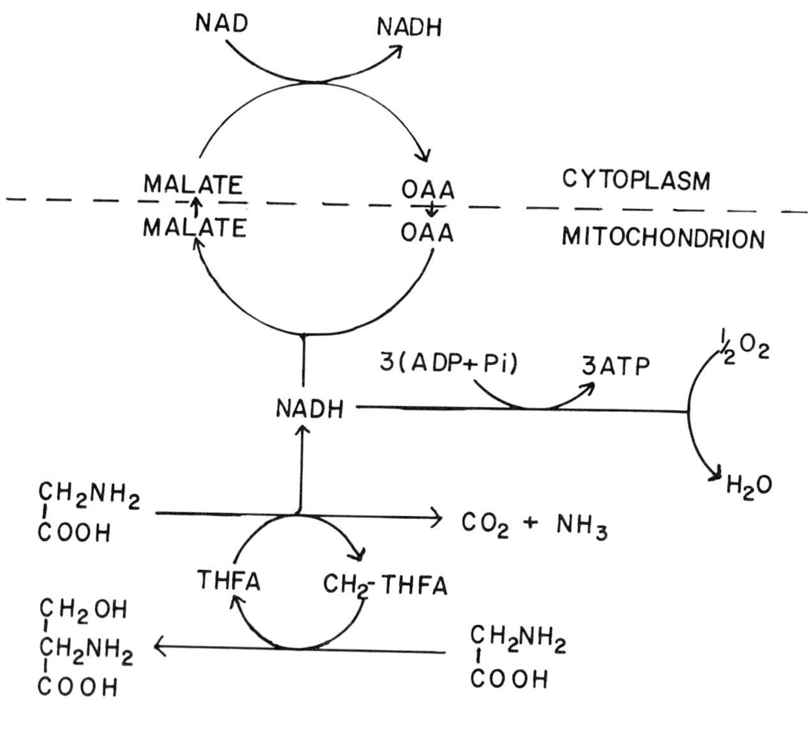

Figure 2. The decarboxylation of glycine and formation of serine in leaf mitochondria.

can be exchanged for OAA in the cytoplasm via the dicarboxylate shuttle and oxidized in the cytoplasm to produce NADH for cytoplasmic reactions. One such reaction is nitrate reduction, and it has been shown with a reconstituted system of supernatant containing malate dehydrogenase and nitrate reductase and mitochondria that glycine oxidation can support nitrate reduction via a malate-OAA shuttle (32). *In vivo* the proportion of NADH that is oxidized by either method is unknown but the high ATP/ADP ratio of the cytoplasm (12) would probably severely limit its oxidation through the respiratory electron transport chain. If this is so, then most of the reducing equivalents from glycine oxidation would be exported from the mitochondrion as malate. As stated earlier these reducing equivalents could be used in the cytoplasm but they could also be further transported to the peroxisome where they could be used for the reduction of hydroxypyruvate. With this linkage, the reduction of NAD with its oxidation in the photorespiratory cycle is balanced and outside control on any one step is avoided. Isoenzymes of malate dehydrogenase are present in the mitochondria, the cytoplasm, the peroxisome and the chloroplast (21) so that a malate-OAA shuttle between all these regions is at least feasible. The rates of glycine oxidation supported by the reduction of OAA are much greater than those supported by the oxidation of NADH through the respiratory electron transport chain and are more consistent with those required for photorespiration (28, 29).

With isolated mitochondria equal amounts of CO_2 and NH_3 are released (Table 1) during glycine decarboxylation. For every glycine decarboxylated a serine is produced (5, 14, 29). If a similar stoichiometry obtains *in vivo* what are the rates of ammonia production that might be expected?

In experiments with leaf discs and $^{14}CO_2$ feeding and chase treatments, the results show that there is a very rapid turnover of glycine and serine (Fig. 3). In 15 min the specific activity of these compounds reaches 60% of the specific activity of the $^{14}CO_2$ supplied. In comparison the

Table 1. Ammonia and CO_2 release from glycine by spinach leaf mitochondria.

	CO_2 Release	Glycine NH_3 Release	Total NH_3 Release	NH_3/CO_2 ratio from Glycine
Data of Keys *et al.* (13)	μmol	μmol	μmol	
Control	0.62	0.63	0.87	1.01
Control + 2-oxoglutarate	0.62		0.87	
Control + glutamate	1.04	1.15	1.31	1.1
Control + glutamate + GS[a]	1.85	0.24	0.35	0.13
Data of Woo *et al.* (30)	$\mu mol\ mg^{-1} chl\ hr^{-1}$			
Control (ADP system)	7.5	7.0		0.93
+ Malate + 2-oxoglutarate	7.3	7.2		0.99
Control (OAA system)	16.0	16.0		1.0
+ Malate + 2-oxoglutarate	13.6	13.4		1.0

[a] Glutamine synthetase.

relative specific activity of 3-phosphoglyceric acid was 85% (17). Glycine and serine were always very similar in their specific activities and similar to the specific activity of the CO_2 evolved. Experiments with ^{18}O and spinach leaves show a saturation of ^{18}O label in glycine within 3 min of feeding and also indicate that the ^{18}O

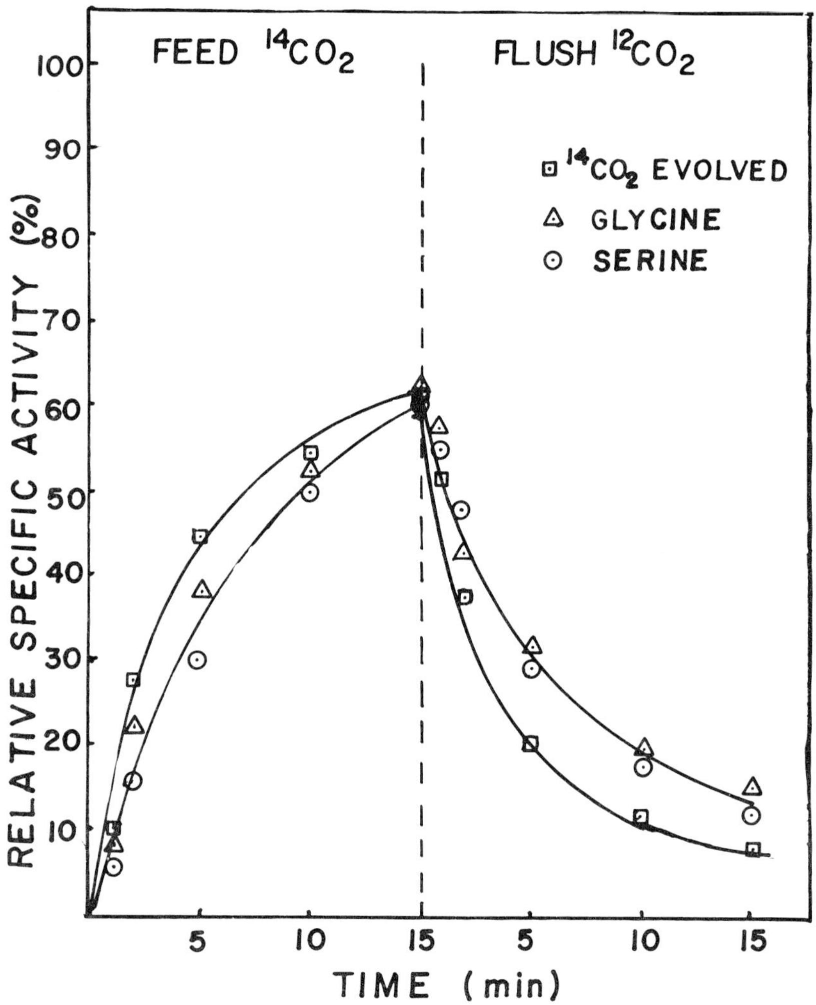

Figure 3. The specific activity (expressed as a percentage of the specific activity of the supplied $^{14}CO_2$) of glycine, serine and the respired CO_2 from sunflower leaves during a $^{14}CO_2$ feeding and $^{12}CO_2$ flushing period. (Data of Mahon et al. (17).)

label is rapidly chased from glycine when ^{16}O replaces the labelled oxygen (3).

From experiments such as those shown in Fig. 3 the flux of carbon through glycine can be calculated. Assuming that one-half of the glycine is decarboxylated the rate of ammonia production by the process can be determined (Table 2). Ammonia production in the three estimates ranged from 24 to 39 μmoles mg^{-1}Chl hr^{-1}. The OAA supported glycine decarboxylation system (Table 1) will sustain these rates, if the activity measured in isolated mitochondria is corrected for 75% breakage using soluble fumarase activity (28). On a fresh weight basis ammonia production by photorespiration is 10 fold the rate of production of ammonia from NO_3^- and several times that determined from NO_2^- (Table 2). It would follow that the capacity of the ammonia assimilation system in C_3 leaves would quantitatively have to be geared to rates of photorespiratory NH_3 release rather than to those of NH_3 formation from primary N-assimilation.

Table 2. Ammonia production by photorespiration and inorganic nitrogen assimilation in leaves.

Source	NH_3 Production		Reference
	μmol g^{-1} F.W. hr^{-1}	μmol mg^{-1} Chl hr^{-1}	
Photorespiration[a]			
	35	24	(17)
		39	(3)
	80		(13)
Nitrate Reduction			
	2-3		(6)
	4-6		(11)
Nitrite Reduction			
	12		(18)
	4		(6)

[a] Ammonia production in photorespiration was calculated from the rate of glycine flux assuming that one-half of the glycine was decarboxylated.

The reassimilation of the NH_3 released in photorespiration could occur through either or both of the two mechanisms available for primary nitrogen assimilation. The two mechanisms available are glutamate dehydrogenase (GDH) and glutamine synthetase (GS) coupled to glutamate (amide):2-oxoglutarate aminotransferase (GOGAT). Both GDH and GS are present in sufficient quantity in wheat leaves to catalyze the rates of ammonia assimilation required (13).

GLUTAMATE DEHYDROGENASE

$$\text{2-OXOGLUTARATE} + NH_3 + \text{NADH} \longrightarrow \text{GLUTAMATE} + \text{NAD}^+$$

GLUTAMINE SYNTHETASE

$$\text{GLUTAMATE} + NH_3 + \text{ATP} \longrightarrow \text{GLUTAMINE} + \text{ADP} + \text{Pi}$$

GOGAT

$$\text{2-OXOGLUTARATE} + Fd_{RED} + \text{GLUTAMINE} \longrightarrow \text{2 GLUTAMATE} + Fd_{OX}$$

Glutamic dehydrogenase is a ubiquitous enzyme in plants and is active with either NAD or NADP. It is predominantly located in the mitochondria (16) and would thus appear to be in the most favourable position to use the ammonia released from the glycine decarboxylation. This reaction would seem to be further privileged because glycine decarboxylation produces NH_3 and NADH in equal amounts, the exact requirements for GDH to catalyze the synthesis of glutamate from 2-oxoglutarate. These apparent advantages of GDH for the reassimilation of photorespiratory NH_3 are, however, completely nullified by the high Km for ammonia. The enzymes from higher plants have Km's for ammonia ranging from 9 to 70 mM (9) and it is unlikely that these concentrations occur. Evidence that NH_3 is not reassimilated by this means in isolated mitochondria is presented in Table 1. The presence of 2-oxoglutarate with or without malate, to provide additional reducing power, does not decrease the amount of ammonia released. These experiments were conducted in air where the electron transport chain could operate and the NADH be reoxidized via O_2. Thus a lack of NADH may be limiting the activity of glutamate dehydrogenase. Under anaerobic conditions, however, where a lack of NAD^+ severely restricts glycine decarboxylation by mitochondria the addition of 2-oxoglutarate did not result in any stimulation of glycine oxidation, suggesting that NADH was not reoxidized by glutamate dehydrogenase (Canvin, unpublished results). Similarly, when the electron transport chain was inhibited with Antimycin A no stimulation of glycine decarboxylation was observed upon the addition of 2-oxoglutarate unless the reaction mixture was supplemented with 16.4 mM $(NH_4)_2SO_4$ (27). As papers from the Rothamsted group (13, 19, 27) have forcefully argued it is most unlikely that GDH functions in the primary assimilation of NH_3 or in the reassimilation of photorespiratory NH_3.

If GDH does not catalyze the reassimilation of photorespiratory NH_3 then we are left with the alternative, glutamine synthetase. In contrast to the poor affinity of GDH for ammonia, glutamine synthetase has a Km for ammonia of 10 to 40 μM (16). When glutamate, glutamine synthetase and ATP are added to the isolated mitochondrial systems there is a considerable reduction in the amount of ammonia release (Table 1) and this represents the reassimilation of the NH_3 (13). Decarboxylation of glycine was also stimulated as measured by $^{14}CO_2$ release (Table 1) and this was consistent with increased ADP production from ATP through the activity of GS (13). This result shows that in a reconstituted *in vitro* system GS can convert the free ammonia back into organic form but is this the mechanism that prevails *in vivo*?

Studies by Woo *et al.* (30) with spinach leaf discs had shown that ^{15}N-ammonia was rapidly incorporated into the amide and amino groups of glutamate and glycine. The results were consistent with a rapid turnover of the amino group of glycine and the incorporation of recently assimilated NH_3 into that position. The high label content of the amide group of glutamine and similar label content between the amide and amino groups suggested that the NH_3 was initially assimilated through the GS/GOGAT system.

More direct *in vivo* evidence for this means of reassimilation of photorespiratory NH_3 was provided by Keys *et al.* (13). These authors incubated wheat leaves or pea protoplasts with ^{15}N-glycine and determined the amount of ^{15}N-NH_3 release (Table 3). Parallel incubations were performed in the presence of methionine sulfoximine (MSO) a specific inhibitor of GS. In the presence of the inhibitor both the total NH_3 release and the glycine NH_3 release increased (Table 3). This evidence strongly suggests that glutamine synthetase is responsible for reassimilation of ammonia as its inhibition results in increased amounts of free ammonia.

Careful studies by Wallsgrove *et al.* (26, 27) have shown that 60% of the glutamine synthetase of pea and barley leaves is located in the chloroplast and 40% located in the cytosol. As ammonia diffusing from the mitochondrion would initially encounter the cytosolic GS it has been suggested that this GS may be most active in the reassimilation of the NH_3 although the chloroplast enzyme may also be involved (26).

The further conversion of the amide group of glutamine to the 2-amino position of glutamate and thus its introduction into a position where it is available for transfer

Table 3. Ammonia release from glycine in tissues and the effect of methionine sulfoximine (MSO). (Data of Keys et al. (13).)

Treatment	Time	NH_3 Release from Glycine	
		control	MSO
	min.	μmol	μmol
Wheat leaves	45	0.6	2.7
	90	1.4	6.4
Pea leaf protoplast	60	0.08	0.16

back to glycine or a wide number of other transfers unquestionably occurs through the action of GOGAT. A ferredoxin dependent enzyme from pea leaves was first described by Lea and Miflin (15) in 1974 and since then it has been shown to be present in *Lemna* and green and blue-green algae (see 16). The Km's for glutamine, 2-oxoglutarate and ferredoxin are 330 μM, 150 μM and 2 μM respectively (25). The enzyme is entirely located in the chloroplast (26) so both glutamine and 2-oxoglutarate, generated in the cytosol or peroxisomes, must be returned to the chloroplast for this reaction. Intact chloroplasts actively catalyze a light dependent assimilation of ammonia (1, 2). The concentrations of ammonia for maximal activity (2) and the inhibition of the process by methionine sulfoximine and azaserine were consistent with ammonia assimilation via the GS/GOGAT mechanism and not the GDH mechanism (1, 2).

Direct and definitive evidence for the role of GS/GOGAT in the reassimilation of photorespiratory NH_3 is provided by C.R. Somerville and W.L. Ogren (23) with studies on mutants of *Arabidopsis thaliana* (L.) Heynh that are deficient in leaf GOGAT. These mutants are viable and not distinguishable from wild type plants if grown in an atmosphere of 1% CO_2 in air, which suppresses photorespiration. If grown in a normal atmosphere, however, the mutants are not viable. The GOGAT activity of the leaves of the mutants was less than 5% of that of the wild type plants. Normal levels of glutamine synthetase and glutamic dehydrogenase were present. Mutant plants which had been growing in a 1% CO_2 atmosphere were transferred to normal atmospheric conditions in darkness for several hours. Subsequently, they were enclosed in a chamber flushed with a gas stream containing 50% O_2 and 357 μl l^{-1} CO_2 and exposed to light. Photosynthesis was measured continuously and after 20 min it had declined to about 7% of the control rate. Ammonia, glutamate and glutamine levels were determined at time zero and at various times thereafter in wild type and mutant plants. The changes in the level of these compounds after 25 min is shown in Table 4. In the wild type, ammo-

Table 4. Ammonia and amino acid content of leaves of Arabidopsis mutant (CS 113) and wild type plants exposed to light in normal levels of CO_2 and 50% O_2. Results are presented as change in concentration after 25 minutes from time zero. (Data of C.R. Somerville and W.L. Ogren (23).)

Compound	Wild Type	Mutant
	μmol mg^{-1} Chl	
Ammonia	+ 0.14	+ 1.6
Glutamate	+ 0.78	- 0.83
Glutamine	+ 0.71	+ 1.24
Serine	+ .95	+ .03
Alanine	+ .20	- .09

nia concentrations increased slightly and both glutamate and glutamine increased in amounts. Serine and alanine also increase in amount and this reflects the increase in pool size of these compounds upon a transfer from non-photorespiratory conditions to photorespiratory conditions. In the mutant plants, ammonia concentrations increased greatly, glutamate concentrations decreased to a low value and glutamine concentrations increased more than in the control plants. Ammonia would initially be reassimilated by glutamine synthetase and this is reflected by the conversion of almost all of the glutamate to glutamine. Because of the lack of GOGAT the amide group of glutamine is not recycled to produce more glutamate and since there is no acceptor for the glutamine synthetase reaction the ammonia accumulates. These results show quite clearly that the GS/GOGAT system plays a central role in the re-assimilation of photorespiratory NH_3. Indeed, as Somerville and Ogren (23) state, this may be the sole requisite function for leaf GOGAT, as the plants, when grown under conditions which suppress photorespiration, are able to assimilate the required nitrogen from nitrate reduction without the need of this enzyme.

LITERATURE CITED

1. Anderson, J.W. and J. Done, 1977. A polarographic study of glutamate synthase activity in isolated chloroplasts. Plant Physiol. 60: 354-359.
2. Anderson, J.W. and J. Done, 1977. Polarographic study of ammonia assimilation by isolated chloroplasts. Plant Physiol. 60: 504-508.
3. Berry, J.A., C.B. Osmond and G.H. Lorimer, 1978. Fixation of $^{18}O_2$ during photorespiration. Plant Physiol. 62: 954-967.
4. Bird, I.F., M.J. Cornelius, A.J. Keys and C.P. Whittingham, 1972. Adenosine triphosphate synthesis and the natural electron acceptor for synthesis of serine from glycine in leaves. Biochem. J. 128: 191-192.
5. Bird, I.F., M.J. Cornelius, A.J. Keys and C.P. Whittingham, 1972. Oxidation and phosphorylation associated with the conversion of glycine to serine. Phytochem. 11: 1587-1594.
6. Canvin, D.T. and C.A. Atkins, 1974. Nitrate, nitrite, and ammonia assimilation by leaves: Effect of light, carbon dioxide and oxygen. Planta 116: 207-224.
7. Canvin, D.T., N.D.H. Lloyd, H. Fock and K. Przybylla, 1976. Glycine and serine metabolism and photorespiration. *In* CO_2 Metabolism and Plant Productivity, ed. R.H. Burris, C.C. Black. University Park Press, Baltimore, pp. 161-176.
8. Canvin, D.T., J.A. Berry, M.R. Badger, H. Fock and C.B. Osmond, 1980. Oxygen exchange in leaves in the light. Plant Physiol. (in press).
9. Davies, D.D. and A.N. Teixeira, 1975. The synthesis of glutamate and the control of glutamate dehydrogenase in pea mitochondria. Phytochem. 14: 647-656.
10. Grodzinski, B., 1978. Glyoxylate decarboxylation during photorespiration. Planta 144: 31-37.
11. Hageman, R.H., 1979. Integration of nitrogen assimilation in relation to yield. *In* Nitrogen Assimilation of Plants, ed. E.J. Hewitt and C.V. Cutting. Academic Press, New York, pp. 591-611.
12. Heber, U. and K.A. Santarius, 1970. Direct and indirect transfer of ATP and ADP across the chloroplast envelope. Z. Naturforsch. 25b: 718-728.
13. Keys, A.J., I.F. Bird, M.J. Cornelius, P.J. Lea, R.M. Wallsgrove, B.J. Miflin, 1978. Photorespiratory nitrogen cycle. Nature 275: 741-743.
14. Kisaki, T., N. Yoshida and A. Imae, 1971. Glycine decarboxylase and serine formation in spinach leaf mitochondrial preparation with reference to photorespiration. Plant and Cell Physiol. 12: 275-288.
15. Lea, P.J. and B.J. Miflin, 1974. Alternative route for nitrogen assimilation in higher plants. Nature 251: 614-616.
16. Lea, P.J. and B.J. Miflin, 1979. Photosynthetic ammonia assimilation. *In* Encyclopedia of Plant Physiology (New Series), Vol. 6, ed. M. Gibbs and E. Latzko. Springer-Verlag, Berlin, pp. 445-456.

17. Mahon, J.D., H. Fock and D.T. Canvin, 1974. Changes in specific radio-activity of sunflower leaf metabolites during photosynthesis in $^{14}CO_2$ and $^{12}CO_2$ at three concentrations of CO_2. Planta 120: 245-254.
18. Miflin, B.J., 1972. The rôle of light in nitrite reduction; studies with leaf discs. Planta 105: 225-233.
19. Miflin, B.J. and P.J. Lea, 1977. Amino acid metabolism. Ann. Rev. Plant Physiol. 28: 299-329.
20. Rehfeld, D.W. and N.E. Tolbert, 1972. Aminotransferases in peroxisomes from spinach leaves. J. Biol. Chem. 247: 4803-4811.
21. Rocha, V. and I.P. Ting, 1971. Malate dehydrogenases of leaf tissue from *Spinacia oleraceae:* Properties of three isozymes. Arch. Biochem. Biophys. 147: 114-122.
22. Somerville, C.R. and W.L. Ogren, 1980. Photorespiration mutants of *Arabidopsis thaliana* deficient in serine-glyoxylate aminotransferase activity. Proc. Nat'l. Acad. Sci. (U.S.A.) (in press).
23. Somerville, C.R. and W.L. Ogren, 1980. Photosynthesis is inhibited in mutants of *Arabidopsis* deficient in glutamate synthase (GOGAT) activity. Nature (in press).
24. Tolbert, N.E. and R.K. Yamasaki, 1969. Leaf peroxisomes and their relation to photorespiration and photosynthesis. Ann. New York Acad. Sci. 168: 325-340.
25. Wallsgrove, R.M., E. Harel, P.J. Lea and B.J. Miflin, 1977. Studies on glutamate synthase from the leaves of higher plants. J. Expt. Botany 28: 588-596.
26. Wallsgrove, R.M., P.J. Lea and B.J. Miflin, 1979. Distribution of enzymes of nitrogen assimilation within the pea leaf cell. Plant Physiol. 63: 232-236.
27. Wallsgrove, R.M., A.J. Keys, I.F. Bird, M.J. Cornelius, P.J. Lea and B.J. Miflin, 1980. The location of glutamine synthetase in leaf cells and its role in the reassimilation of ammonia released in photorespiration. J. Expt'l. Botany (in press).
28. Woo, K.C. and C.B. Osmond, 1976. Glycine decarboxylation in mitochondria isolated from spinach leaves. Aust. J. Plant Physiol. 3: 771-785.
29. Woo, K.C. and C.B. Osmond, 1977. Participation of leaf mitochondria in the photorespiratory carbon oxidation cycle: Glycine decarboxylation activity in leaf mitochondria from different species and its mitochondrial location. *In* Photosynthetic Organelles, ed. S. Miyachi, S. Katch, Y. Fijita, K. Shibata. Japan Soc. Plant Physiol., Ctr. Academic Publ., Tokyo, pp. 315-323.
30. Woo, K.C., J.A. Berry and G.L. Turner, 1978. Release and refixation of ammonia during photorespiration. Carnegie Inst. Yearbook 77: 240-245.
31. Woo, K.C., 1979. Properties and intramitochondrial localization of serine hydroxymethyltransferase in leaves of higher plants. Plant Physiol. 63: 783-787.

32. Woo, K.C., M. Jokinen, D.T. Canvin, 1980. Reduction of nitrate via a dicarboxylate shuttle in a reconstituted system of supernatant and mitochondria from spinach leaves. Plant Physiol. 65: 433-436.

CHAPTER

SEED STORAGE PROTEINS: - GENETICS, SYNTHESIS, ACCUMULATION AND PROTEIN QUALITY

B.J. Miflin and P.R. Shewry
Biochemistry Department, Rothamsted Experimental Station, Harpenden, Herts., U.K.

Introduction
Characterization
- Cereals
- Legumes
- Immunology
- Subunit Composition
- Carbohydrate Composition
- Amino Acid Composition of Storage Proteins

In Vivo Synthesis
In Vitro Synthesis
- Cereals
- Legumes

Genetics
- Chromosomal Location of Structural Genes
  - Barley
  - Wheat
  - Rye
  - Maize
  - Legumes
- Homology of Storage Polypeptides - Evidence for Duplication and Divergence of Structural Genes
  - Cereals
  - Legumes
- Regulatory Genes which affect Protein Synthesis and Quality
  - Cereals
  - Legumes
- Structural Mutations at Storage Protein Loci

Concluding Discussion and The Way Ahead
- Improvements in Protein Quantity

J.D. Bewley (ed.), Nitrogen and Carbon Metabolism

Improvements in Protein Quality
Changes in the total amount of storage proteins
Changes in the relative proportion of the individual components
Changes in the amino acid composition of the storage protein
Changes in the nature of other nitrogenous fractions in the seeds
Acknowledgements
References

INTRODUCTION

The major primary source of dietary amino acids for non-ruminant animals is the protein present in the harvested grain of crop plants. Of this about half is composed of a fraction termed storage proteins. The major sources of seed protein are shown in Table 1. From this it is obvious that the chief contribution is made by the cereal crops with legumes the second most important source - other crops make a relatively small contribution. This article will deal only with the cereal and legume storage proteins and within this group concentrate on those areas where most is known about the biochemical nature of the proteins and their deposition.

The proteins of seeds were studied as long ago as the 18th century with a considerable amount of work done throughout the 19th century. The first comprehensive reviews of the subject are usually credited to Osborne and the second edition of his book gives a good account of the historical development of the subject (Osborne, 1924). Osborne classified proteins on the basis of their solubility into albumins, globulins, prolamins and glutelins, and his separation techniques are still often used today. Whilst they remain a reasonable framework for protein classification there have been improvements in methodology over the last eighty years and we have reviewed some of them, as they relate to cereals, elsewhere (Miflin and Shewry, 1979b). This classification is however strictly operational and tells us little of the physiological role of the different proteins. From a physiological standpoint proteins may be classified in terms of their function i.e. enzymatic, structural, storage etc. or in terms of their location, e.g. membrane, chloroplast stroma, peroxisomal, protein bodies etc. Thus, with these possible approaches in mind, we consider that some of the properties of seed storage proteins are that:-

(1) Their predominant function is to act as a store of nitrogen and sulphur (and in some cases carbon) that is laid down during seed development and utilized during germination.

Table 1 Storage protein production from grains

Crop	Grain yield in 10^6 MT	% of Protein in grain	Approx. % of in storage proteins	Estimated world production of storage proteins
Cereals	1580			prolamins
Wheat	441	12.2	55	30
Maize	362	9.5	55	19
Rice	376	8.5	-	
Barley	196	11.0	40	9
Rye	32	11.0	-	
Oats	50	13.0	-	
Millet	37	9.7	55	
Sorghum	69	10.1	55	
				Legume globulin
Total Pulses	62	23	65(40)	9.0
Dry beans	17	22.1	65(4)	2.5
Broad beans*	12	23.4		2.0
Peas*	16	22.5		2.3
Chick peas	7	20.1		1.0
Lentils	1	24.2		0.2
Soybeans	80	38.0	65(4)	20

Data taken from (1) FAO statistics for 1978 FAO (1979)

(2) FAO values for food contents FAO (1970)

(3) from table 2

(4) approximate estimate

(2) They accumulate relatively late in seed development.

(3) Their synthesis is preferentially stimulated by increasing the nitrogen supply to the plant.

(4) Generally they consist of proteins chiefly of one solubility class for a given species and consist of a polymorphic series of related polypeptides.

(5) They are specified by a relatively small number of genetic loci, probably complex, each consisting of a number of duplicated sequences.

(6) They are organ specific i.e. only found in the seed storage organs.

(7) They are deposited in protein bodies.

In the remainder of this paper we will present evidence in support of these assertions for the major species of cereals and legumes and discuss how the proteins are important in crop quality and quantity.

CHARACTERIZATION

The names and properties of the major seed storage proteins are given in Tables 2 and 3.

Cereals

Within the cereals there are divisions in the nature of the storage proteins. In the starchy endosperm of wheat, barley, and maize the proteins fall into the class of prolamins; in oats the major group is the globulins and in rice the glutelins, although in both of these latter species some prolamins are present. Besides the storage proteins in the starchy endosperm there are also a small amounts of storage proteins in the aleurone layers and in the scutellum and embryonic axis of most cereals. These are likely to be globulins as Danielsson (1949) has described an 8S globulin located in the embryo.

This classification (Table 2) is unlikely to be accepted by all workers in the field since many still consider that the glutelins form a significant part of the storage proteins of wheat, maize and barley. This viewpoint appears to be most strongly held for wheat (e.g. see Wall, 1979; Kasarda et al., 1977; Bushuk, 1980b). We consider that this view is misguided and that the usually studied glutelin fraction of wheat (see Miflin et al., 1980), maize (see Wilson

Table 2. The distribution of cereal seed N into the major solubility classes

Species	Non-protein N	Albumins	Globulins	Prolamins[a]	Glutelins	Residue	Name of Prolamin	Reference
Hordeum vulgare[b]	11.6	15.6		45.2	18.0	5.2	Hordein	1
Triticum aestivum[c]		33.6		57.5	10.1		Gliadin	2
Zea mays[b]		6.5		64.9	15.2		Zein	3
Sorghum vulgare[b]	8.0		6.2	53.8	25.5	6.8	Kafarin	4
Oryzae sativa[b]		15.7		6.8	61.5	15.4		5
Avena sativa[c]		11.0	56.0	9	23		Avenin	6

(a) The alcohol soluble and alcohol + reducing agent soluble fractions have been added together to give the prolamin fraction even though in some cases the original authors did not consider the latter fraction as a prolamin (see text); (b) as % total seed N; (c) as % recovered seed N.

References

1 Shewry et al., 1978b; 2 Miflin et al., 1980b; 3 Miflin and Shewry, 1977 (Table 3 data of Milan group); 4 Paulis and Wall, 1979; 5 Villareal and Juliano, 1978; 6 Peterson and Smith, 1976.

Table 3 Globulin Storage Proteins of Legumes

Species	Approximate Sedimentation value	Specific name	Subunit mol. wt. on SDS-PAGE (Approximate in kilodaltons)	Reference
Pisum sativum	7S	Vicilin	56, 43, 23	1
..	7S	..	75, 50, 30, 18	2
..	11S	Legumin	40,20	1,2,3
Vicia faba	7S	Vicilin	56, 46, 33, 31	1
..	11S	Legumin	37, 20	4
Phaseolus vulgaris	7S	G-1	53, 47, 43	5
..	7S	Glycoprotein-II	43, 35*	6
Phaseolus aureus	8S	Vicilin	63, 50, 30, 24	7
..	11S	Legumin	56, 44, 16	7
Glycine max	7S (non-dimerizing)	α - conglycinin		8
	7S (dimerizing)	β-conglycinin	57, 42	8
	11S	glycinin	42, 37, 19	9

* determined by ultracentrifugation. References: 1 Derbyshire et al., 1976; 2 Thomson et al, 1978; 3 Casey, 1979a; 4 Wright and Boulter, 1974; 5 Hall et al., 1977; 6 Pusztai and Watt, 1970; 7 Ericson and Chrispeels, 1973; 8 Thanh and Shibasaki 1976a,b; 9 Moreira et al. 1979

et al., 1980), and barley (see Shewry et al., 1978b) are in fact the contaminating residues of the prolamin fraction left behind by non-exhaustive and inadequate extraction techniques. This view is based both upon a careful analysis of the component polypeptides and the amino acid composition of the so-called glutelin fractions.

The cereal storage proteins (excluding oats) are extremely difficult to solubilize and cannot be studied in the absence of denaturing solvents. It is thus not easy to determine if any specific multimeric complexes exist although prolamins do form disulphide linked aggregates. Because of this it is the component polypeptides of the protein fraction that have been most studied. Various techniques based on electrophoresis have been used to separate these polypeptides particularly for analytical procedures; these include starch gel electrophoresis (SGE), sodium dodecylsulphate polyacrylamide gel electrophoresis (SDS-PAGE), urea-PAGE, aluminium lactate-PAGE, isoelectric focusing (IEF), and two-dimensional procedures based on IEF (Wrigley and Shepherd, 1973; Shewry et al., 1978a), or urea-PAGE (Shewry et al., 1978a) in the first dimension and SDS-PAGE or SGE (Wrigley and Shepherd, 1973) in the second dimension or on changing pH between dimensions using aluminium lactate-PAGE (Mecham et al., 1978). Examples of some of the separations that may be achieved are shown in Fig. 1.

The two-dimensional separations have extremely high resolving power and demonstrate the important principle that one-dimensional gels, especially SDS-PAGE, lead to a gross over-simplification. The excessive reliance on SDS-PAGE as a separation system, particularly in research with maize, has led to some controversy and to a fundamental misconception of the nature of the storage proteins. Thus it has been considered that zein consists of two major components of about 22 and 19 kd (kilodaltons) which might be specified by two mRNAs (e.g. see Larkins et al., 1976a,b; Burr et al., 1978) and that the complexity and microheterogeneity revealed by IEF could be due to hydrophobic aggregation, disulphide bridging, random deamidation, incorrect proteolytic cleavage or glycosylation during processing (Burr, B. 1979; Larkins et al., 1980). Despite these assertions almost all of the evidence that is available supports the concept that the cereal seed proteins consist of a polymorphic series of proteins. This evidence is briefly

(1) That attempts to show that the microheterogeneity is due to amidation artefacts have failed. (Righetti et al., 1977; Gianazza et al., 1976).

(2) That microheterogeneity can be revealed by systems not involving IEF e.g. urea-PAGE (Alexandrescu et al., 1976; Paulis and Wall, 1977) and two-dimensional urea PAGE/SDS-PAGE (Shewry et al., 1978a).

(3) That the IEF bands are inherited in a manner that is subject to gene dosage (Soave et al., 1979) and that the genes specifying the bands are distributed on different maize chromosomes (Soave et al., 1979, 1980; Valentini et al., 1978).

(4) That the IEF bands are produced by the translation of cereal storage protein mRNA and polysomes in in vitro systems (Viotti et al., 1978, 1979; Miflin and Shewry, 1979a; Miflin et al., 1980c)

(5) Studies on nucleic acid hybridization in vitro and in situ suggest a multiplicity of zein genes (Viotti et al., 1979, 1980a, ; Soave et al., 1980 but see Larkins et al., 1980 for an alternative view).

(6) Sequencing studies of cereal protein reveals that microheterogeneity in amino acid sequences does exist (Kasarda, 1980).

It is therefore our conclusion that the multiplicity of polypeptides seen in the two-dimensional analysis of cereal prolamins is real and reflects the fact that they are an extremely polymorphic group of proteins (similar polymorphism also exists in animal storage proteins e.g. see Thomson, 1975). Further, because IEF only reveals substitutions involving amino acids of different charge, which probably only represents a minor proportion of the potential heterogeneity (there being more uncharged amino acid residues that charged ones), the number of polypeptides revealed by 2-D separations represents a minimum rather than a maximum.

Legumes

It is almost universally accepted that the storage proteins of legume seeds are globulins. The claim that some albumins may act as storage proteins (Murray, 1979) is not in our view substantiated because as yet there is virtually no evidence that any of the properties referred to above are met. The controversy or confusion in the legume field appears to arise when the proteins are subdivided. Osborne and Campbell (1898) originally separated two protein fractions from the globulins of pea seeds, the fraction that was more soluble in dilute salt solutions and which was coagulated by heat (90°C) they termed vicilin and that fraction less soluble and not heat coagulable they termed legumin. Subsequently Danielsson (1949) separated globulins from pea seeds by means of ammonium sulphate fractionation and isoelectric precipitation at pH 4.7 and showed they were comparable with legumin and vicilin. By means of sedimentation studies in the ultracentrifuge he showed that they each migrated as single components, vicilin with a sedimentation coefficient (S value) of around 7 and legumin around 11. In addition the studies showed that a wide range of legume seeds contain protein species of about 7S and 11S and on this basis he termed them vicilins and legumins. However, the various legume seed globulins have also been given specific names. Some examples of nomenclature and characteristics of different legume seed globulins are given in Table 3. Although many workers use the general terms vicilin and legumin others specify the sedimentation coefficients of the proteins. Sedimentation is however not a invariable characteristic and depends upon both the pH and the ionic strength of the solvent. Thus 11S proteins can dissociate to 7S and then to 3S and 2S. Furthermore, considerable variation exists in the reported values for the same globulin (see Derbyshire et al., 1976 for further discussion). Also confusing results may be obtained with 7S vicilin-like proteins; thus the conglycinins have been divided into α and β fractions dependent on whether they dimerize or not (Thanh and Shibasaki, 1976a). However recent work has shown that one of the holoproteins (β_3) of the β-conglycinins does not form the normal dimer (Gayler and Sykes, 1980). Hall and coworkers (McLeester et al., 1973; Sun and Hall, 1975; Hall et al., 1980) have argued against overgeneralization of nomenclature pointing out that their G1 fraction from *Phaseolus vulgaris* sediments at about 7S at neutral pH (i.e. like vicilin) but that it requires stronger concentrations of salt to be solublized (i.e. legumin like) than the G2 fraction. The G2 fraction also sediments at 7S at neutral and alkaline pH. In many respects we would agree with the general view that at present the legume storage proteins are poorly defined. However, careful purification and characterization of pea legumin (Casey, 1979a) and soybean conglycinin (Thanh and Shibasaki, 1976a,b, 1977, 1978a,b) show that certain systems are now being far better defined than in the past.

Immunology

One approach to characterization and the determination of relationships is to use immunological techniques. Vicilin and legumin have been injected separately into rabbits and antibodies raised which show no cross reactivity. Using such antibodies Dudman and Millerd (1975) have tested storage protein from several tribes of the Leguminosae and found that the Trifolieae and the Fabeae contain proteins antigenically related to legumin or to vicilin from Vicia faba. However the Phaseoleae gave no cross reaction with either of the antibodies. Thus indicating a clear division within the Leguminoseae with regards to this property although sequencing evidence suggests some similarities.

Subunit composition

Studies on subunit composition have chiefly utilized SDS-PAGE under reducing conditions. From this it has been shown that the legume storage protein consists of a number of individual polypeptides of different molecular weights (Table 3), for example pea and broad bean seed legumins have acidic subunits of around 30-40 kb and basic subunits of about 20kb. However, as in cereals the bands obtained on one-dimensional SDS-PAGE do not represent single polypeptides. Some of the first IEF studies by Catsimpoolas (1969) and Catsimpoolas and Wang (1971) showed considerable microheterogeneity of the components of glycinin. Recent work by Casey (1979a,b) and by Gatehouse et al. (1980) has utilized 2-D separations and demonstrated that all the SDS-PAGE bands in legumin are composed of multiple polypeptides.

The nature of the component polypeptides can also be investigated using immunological techniques. In the protein bodies of mature pea seeds there are seven antigenically distinct protein families, one of which corresponds to legumin and four to vicilins; the other two were probably not related to either of the main groups of storage proteins (Millerd et al., 1978). All of the vicilin families contained peptides of the 50 and 30 kd groups but those of the 75 kd were only found in one. Although no firm conclusion about the family relationships among the vicilin polypeptides can be drawn from this, or a related study with Vicia (Manteuffel and Schotz, 1975), it appears that a considerable degree of heterogeneity exists within the vicilin fraction.

Models of the structure of the holoproteins from the various subunits have been proposed. Thus Wright and Boulter (1974) have proposed that legumin of P. sativum (Mol. Wt. 388-410 kd) is composed of 6 basic (Mol. wt. 40 kd each) and 6 acid (Mol. wt. 20 kd) subunits with one acidic and one basic subunit forming a disulphide bonded pair and then the 6 pairs forming the complete molecule. Badley et al. (1975) have proposed, on the basis of electron microscope studies of glycinin, that the 12 sub-units are packed into hexagons placed one on top of the other to form a hollow oblate cylinder. However this may be an oversimplication for although legumin has generally been regarded as a single protein species (Manteuffel and Schotz, 1975; Millerd et al., 1978), because there is now known to be microheterogeneity in subunits, there may be a similar family of closely related, barely distinguishable legumins (see also Moreira et al., 1979). In contrast it has been widely recognized (see above) that vicilin in P. sativum and V. faba is comprised of several different proteins. Recently Thanh and Shibasaki (1976a,b; 1977, 1978a,b) have provided convincing evidence that this is the case for the 7S proteins from soybean. They have isolated six 7S proteins termed ß conglycinins B_1 - B_6 and showed that they are each comprised of three subunits. These subunits can be separated on gels and are termed α (mol. wt. 68 kd), α^1 (mol. wt. 68 kd) and ß (mol. wt. 42 kd) and they come together in different combinations to give the different 7S holoproteins thus B_1 is α^1, β_2, B2, $\alpha\ \beta_2$; B3, $\alpha\ \alpha_1^1\beta$; B4, $\alpha_2\beta$; B5 $\alpha_2\alpha^1$ and B6 α_3. In all one would predict ten possible combinations so that four other possibilities may also exist. At least one of these has now been found in that Sykes and Gayler (1980) report that the β_3 trimer is present in soybean cotyledons. The α, α^1 and ß subunits differ slightly in their amino acid compositions as indicated in Table 4 with the ß subunit being particularly deficient in S amino acids. The α and α^1 subunits are immunologically cross reactive but distinct from the ß subunit. Once again the picture is further complicated when the subunits are subject to IEF as further microheterogeneity is revealed (Thanh and Shibasaki, 1976b, 1977, 1978a). The 11S protein of soybean, glycinin, has also been intensively studied. This protein resembles legumin in that it is composed of disulphide-linked acidic and basic subunits. Kitamura and Shibasaki (1975, 1977) have purified four acidic subunits (A1-A4). Three of the subunits (A1-A3) had a mol. wt. of 37 kd, were closely related antigenically.

Table 4 Partial amino acid composition (in mole %) of certain storage proteins

	Barley		Wheat			Soybean			Pisum	
	Hordeins		Gliadins			7S		11S	7S	11S
Amino Acid	B	C	α_1	β	ω	α subunits	β	basic subunits		
Lys	0.5	0.2	1.4	0.5	0.5	6.2	5.3	3.14	7.9	4.9
Arg	2.4	0.8	3.0	1.4	1.2	8.7	7.2	5.4	7.3	10.5
His	2.1	1.1	1.8	1.1	0.8	1.2	2.0	1.5	2.1	2.8
Asx	1.4	1.0	3.5	2.5	0.6	11.8	13.0	12.9	12.0	12.5
Glx	35.4	41.2	31.7	37.4	51.5	20.5	16.7	14.4	19.3	21.1
Pro	20.6	30.6	12.9	17.5	21.2	7.1	5.1	5.9	3.5	4.3
Cys	2.5	tr	3.1	2.4	0.1	tr	0	1.6	0.35	0.71
Met	0.6	0.2	1.3	0.6	0.1	0.4	0.1	1.1	0.22	0.65
Reference	1	1	2	2	2	3	3	4	5	5

(1) Shewry et al. (1980d) (2) Charbonnier (1973) (3) Thanh and Shibasaki (1977) (4) Okubo et al. (1969) (5) Boulter (1980)

The fourth subunit (A4) had a mol. wt. of 45 kd and was serologically distinct.

Carbohydrate Composition

It is generally considered that the storage proteins of legumes are glycoproteins (e.g. see Derbyshire et al., 1976) whereas those of the cereals are not. Recently this position has been challenged. In cereals Burr et al. (1978) and Burr, F.A. (1979) have reported that zein is a glycoprotein based upon periodate-Schiff staining of zein bands on SDS-PAGE. However this technique is not totally reliable and no evidence was presented that the carbohydrate remains associated with the protein through subsequent purification.

In legumes Casey (1979a) and Gatehouse et al., (1980) have both reported that legumin purified from mature, dried pea seeds does not contain carbohydrate. In both studies the methodology appears to have been carefully applied and the results to be conclusive for the samples studied. Similarly Koshiyama and Fukushima (1976) conclude that glycinin is not glycosylated and Bailey and Boulter (1970) reported only very low levels of carbohydrate associated with V. faba legumin. In contrast Basha and Beevers (1976) have reported that legumin contains 1% neutral sugars and 0.1% amino sugars. Davey and Dudman (1979) have also found carbohydrate associated with legumin and vicilin. The latter authors passed these two storage proteins, separately, over Con-A Sepharose and found that part of the protein bound to the column. The bound portion contained the majority of the carbohydrate associated with the proteins and a proportion of several of the polypeptides. No clear-cut distinction could be made between subunits which bound and those that did not. When periodate-Schiff stained gels were scanned the majority of the stain was associated with low molecular weight polypeptides in the vicilin fraction and with both the 40 and 20 kd legumin subunits. Similar association of carbohydrate with only the low mol. wt. bands of vicilin preparations have been shown by Gatehouse et al. (1980) who consider such bands to be contaminants. This is not unlikely when it is known that pea phytohaemagglutins are associated with the storage globulins and have mol. wts. of 17 kd and less (Millerd et al., 1978; Trowbridge, 1974). The 7S proteins from soybean (ß conglycinin) do appear to be glycosylated and Yamauchi et al. (1975, 1976) have defined the amino acid sequence around the glycosylation site. They isolated, by pronase digestion, tripeptides containing carbohydrate from purified 7S proteins from soybeans.

Two sequences were found Asn-Gly-Thr and Asn-Ala-Thr with N-acetylglucosamine bound to the asparagine residue; the remainder of the carbohydrate was a further N-acetyl glucosamine with a chain of 7-9 mannose units attached. Based on a mol. wt. of 180-210 kd for the 7S protein with 5% carbohydrate, and a mol. wt. for the carbohydrate chain of 1.6-1.9 kd, there would be 4.7 to 6.7 carbohydrate chains per holoprotein. This suggests that there might be 1 carbohydrate chain per subunit, a value supported by analysis of the individual subunits (Thanh and Shibasaki, 1977). The glycoprotein nature of Phaseolus proteins has also been carefully investigated. Pusztai and Watt (1970) report values of 1.0% glucosamine and 3.2% mannose for purified glycoprotein II of P. vulgaris. Ericson and Chrispeels (1976) report lower values for P. aureus 7S protein of 0.2% glucosamine and 1% mannose, the carbohydrate chain is linked to asparagine and is 12 residues long (2 N-acetylglucosamine, 10 mannose) with alanine and threonine proposed as the next residues to asparagine. The amount of carbohydrate is thus only sufficient for 1 chain per holoprotein which they propose consists of 4 subunits of 50 kd, only one of which would be glycosylated.

In summary therefore it has been established that certain legume storage proteins may contain asparagine which has a carbohydrate chain attached via an N-acetylglucosamine and then a chain of one further N-acetylglucosamine plus about 7-10 mannose residues. The sequence of amino acids adjacent to asparagine and the nature of the glycoproteins bear great similarities to those described for animal proteins (Marshall, 1972). However, not all 7S (vicilin type) subunits in a species may be glycosylated and there is good evidence that the 11S (legumin-type) protein may be non-glycosylated in some instances. One possible reason for these variations will be discussed later.

Amino Acid Composition of Storage Proteins

The partial amino acid composition of certain storage proteins is given in Table 4. Particular points of interest are that; the Glx and Asx residues are almost always highly amidated; the proportion of arginine is large in legume proteins but small in cereals which, in contrast, have very large amounts of glutamine and proline; certain of the components of storage proteins have almost no S-amino acids (e.g. the ω gliadins, the C hordeins and the 7S proteins (or vicilins) from legumes, and that cereal proteins are low in lysine and other basic amino acids. Although bulk P. sativum vicilin fractions may contain some S-amino acids labelling studies with seeds suggest that almost all of the ^{35}S was located in one of the 50 kd subunits (Millerd et al., 1979). Similarly Moreira et al. (1979) have shown several-fold variation in the methionine content of individual glycinin subunits. From the data amassed a general conclusion may be drawn, that within each species, the storage proteins fall into two types, those that contain N and S and those that store chiefly or solely N.

IN VIVO SYNTHESIS

Many studies have shown that the synthesis of storage proteins in cereals and legumes takes place relatively late in seed development such that little is formed during the first 1-2 weeks after fertilization followed by a period of rapid accumulation continuing through until maturation. Within the cereals there is little evidence of major differences in the rate of synthesis of the different polypeptides although studies with barley suggest that the C hordeins may form a greater proportion of the total prolamins at the earlier stages of endosperm development (Shewry et al., 1979b). In peas and soybeans relatively detailed studies have been made on the development of the 7S and 11S proteins and of their constituent subunits. These suggest that vicilin appears before legumin in *P. sativum* and *V. faba* cotyledons (Wright and Boulter, 1972; Millerd et al., 1978); however sensitive immunological techniques have shown that legumin is present at very early stages when the embryo only weighs 2 mg (Domoney et al., 1980). Within the constituent polypeptides of the 7S but not 11S fraction differences exist in the relative rate of formation (Wright and Boulter, 1972; Gayler and Sykes, 1980). The reasons for the differences in temporal development are as yet unknown. In particular it is not known if there is a genetically determined programme of synthesis and if so what is the nature of the controlling genes, and also if the sequence is in any way affected by the nutritional status of the seed. The development of *in vitro* culture techniques for legume cotyledons from the earliest stages of development (Stafford and Davies, 1978) may provide a system for further investigation.

As yet insufficient information is available to discriminate between environmental and genetic effects - e.g. how much variation in subunit synthesis depends on the relative supply of N and S and how much depends of specific genetic controls. Further the nature of the factors that regulate the amount of synthesis of the polypeptides *in vivo* is not clear - thus although we know that extra N leads to the synthesis of more total storage proteins we have no knowledge of what governs the relative amounts of component polypeptides synthesized. Part of the answer must be genetic since it is known that the ratio of legumin to vicilin in peas is genetically controlled but whether this involves specific regulatory genes or is solely dependent on the relative degree of gene duplication of the structural loci has not yet been tested. Similarly the most likely reason for the variation in amounts of different cereal polypeptides may reside in the degree of structural gene reiteration. One control that obviously exists is that regulating overall expression of storage proteins which appear to be totally organ specific, however the mechanism whereby this is achieved is unknown and much interesting work in this area of storage protein synthesis remains to be done.

Studies on developing seeds generally suggest that the storage proteins are stored in protein bodies, however, there is no clear consensus as to the origin and nature of the protein bodies or their diversity. The most general view, based on protein body formation in animal tissue, is that the proteins are synthesized on the rough endoplasmic reticulum (RER), pass into the lumen and then to the Golgi apparatus, from where they are cut off in vesicles which transport them to larger vesicles (or vacuoles) where they are deposited. This view of events has been proposed for protein bodies of oil seeds (Dieckert and Dieckert, 1976) and legumes (Boulter, 1979). The role of the Golgi has been questioned (Bain and Mercer, 1966; Neumann and Weber, 1978) and it is difficult to find conclusive evidence for their involvement. A modification of this theory is that the vesicles are budded off directly from the RER (Matile, 1976). In both cases the site of protein synthesis (RER) is spatially separated from the site of protein deposition (vacuole).

A second hypothesis of long standing is that protein bodies are independent organelles capable of both synthesis and deposition of the storage protein and are perhaps of plastid origin. The most clear exposition of this theory is proposed by Morton et al. (1964). They found that wheat protein bodies contained RNA and ribosomes within a distinctive lipoprotein membrane and were capable of in vitro protein synthesis. Later studies by Wilson (1966) inferred that the latter observation may have been due to bacterial contamination. Recently it has been shown that polysomes and mRNA isolated from maize protein bodies can specify the synthesis of storage protein, however, electron micrographs show that these ribosomes are attached to the outside of the protein body membrane (Burr et al., 1978; Larkins and Hurkman, 1978; Viotti et al., 1978).

A third hypothesis is that the protein bodies are formed directly by the synthesis of storage protein on the RER and its deposition into the lumen where it aggregates and accumulates. This has been suggested for maize (Khoo and Wolfe, 1970; Larkins and Hurkman, 1978) and for barley (Miflin and Shewry, 1979a). Both this theory and the plastid theory propose that the sites of synthesis and deposition are not spatially separated and thus transport of the proteins is not required.

Although there is general agreement that the first theory applies to legumes, attempts to reach general conclusions for cereals are not easy. Thus, whilst some authors consider all protein bodies to be of vacuolar origin (Matile, 1976; Altschul et al., 1966; Ashton, 1976; Pernollet, 1978) and it is suggested that hordeins are deposited in protein bodies surrounded by a tonoplast (Ingversen, 1975; Munck and von Wettstein, 1976; Cameron-Mills et al., 1978) others including ourselves, dissent from this view. To try to resolve the issue we have recently carried out a series of experiments isolating and characterizing the subcellular organelles in developing storage tissues (Miflin et al., 1980a,c). The results have shown a clear division between the situation in legumes and cereals. The developing protein bodies of pea cotyledons are surrounded by a membrane. This has none of the characteristics of RER and the bodies co-sediment with marker enzymes (N-acetylglucosaminidase and mannosidase) considered characteristic of vacuoles (see also Wilden et al., 1979). In contrast the protein bodies of cereals have NADH-cytochrome c reductase and RNA (but not vacuolar marker enzymes) associated with them. Evidence for association of the cereal protein with the major peak of RER was also found. Some differences were found between protein bodies of maize and those of wheat and barley; the former had far more RER associated with them and were protected from protease attack whereas the latter had much less RER, did not appear to be surrounded by a complete single membrane and were subject to attack by protease. Our current hypothesis is thus that in all three cereals the storage proteins are formed on the RER pass into the lumen and aggregate in clumps within it. In wheat and barley these clumps break away from the RER and form deposits (protein bodies) with small amounts of RER still attached. In maize, however, the protein aggregates into spherical deposits that accumulate in the RER which completely surrounds them. The reasons for the differences are not yet clear but may lie either in the difference in the properties of the RER or in the known differences in the nature of the storage proteins.

The results also emphasize that, in cereals, the major movement of the storage protein is across the RER whereas in legumes, besides this translocation, it is also necessary for transport of the proteins to the vacuoles to occur. In terms of the theory of protein topogenesis expounded recently by Blobel (1980) one would expect more processing events and signal sequences to be involved in legume storage protein synthesis than in cereals. In particular, based on parellels in other systems, it might be expected that the legume proteins might be glycosylated during or prior to their transportation. However our results show that their final site of deposition occurs in an organelle that is also accumulating the enzymes of deglycosylation, which are known to be important during the breakdown of storage proteins during germination. Consequently, depending on the local environment, the structure of the proteins, rates of dehydration etc. it may be that variable amounts of deglycosylation occur prior to maturation and this may well explain the variable results relating to the glycosylation of legumin and vicilin that exist in the literature (see above). Aspects of the processing of storage proteins will be further discussed below.

IN VITRO SYNTHESIS

Cereals

Studies with maize and barley have shown that when membrane bound polysomes are isolated from developing endosperms and incubated with a wheat germ translation system the products resemble the authentic storage proteins (Larkins and Dalby, 1975; Brandt and Ingversen, 1976; Fox et al., 1977; Larkins and Hurkman, 1978; Miflin and Shewry, 1979b; Matthews and Miflin, 1980).

The polysomes associated with the outside of maize protein bodies also give the same results (Burr and Burr, 1976; Burr et al., 1978, Viotti et al., 1978). When poly A^+ mRNA has been purified, usually from membrane bound polysomes, it too leads to the synthesis of prolamin-like polypeptides. However, poly A^+ directed protein synthesis produces polypeptides that move significantly slower on SDS-PAGE (and are thus presumed to be larger)(Burr et al., 1978; Weinand and Feix, 1978; Brandt and Ingversen, 1978; Larkins and Hurkman, 1978; Miflin et al., 1980c; Matthews and Miflin, 1980). Similar, larger products are also observed with barley polysomes that are incubated with a wheat germ preparation capable of reinitiation (Miflin et al.,

1980c; Matthews and Miflin, 1980). In the barley system the larger products still have the same pI as authentic hordeins. These results have been interpreted in terms of the signal hypothesis (Blobel and Dobberstein, 1975a,b; Blobel et al., 1979) in which proteins made on the rough endoplasmic reticulum (RER) and transported into the lumen have an extra sequence which is uncharged and predominantly composed of hydrophobic amino acids. This sequence acts as a signal in that it attaches the nascent polypeptide chain and the ribosomes to the ER and passes through the ER to the lumen with the rest of the polypeptide chain following. The signal sequence is usually at the N-terminus and is cleaved off after passage of the protein through the membrane (but not always cf. ovalbumin Lingappa et al., 1979). The transport and processing for RER synthesised protein occurs co-translationally and has been demonstrated in vitro for animal (Blobel et al., 1979) and plant (J.M. Lord personal communication) messages using dog pancreas microsomes in a wheat germ system. When the products of translation are transported into the microsomal vesicles they are protected from protease attack. Correct processing of poly A^+ RNAs of a number of RER synthesized animal proteins has been successfully achieved by injecting them into Xenopus laevis oocytes (Lane et al., 1980). Evidence that the processing of cereal protein messages is directly analogous to the animal system has recently been obtained. Larkins et al. (1979) have shown that when zein mRNA is injected into Xenopus oocytes the translation products are of the authentic size and are present in vesicles. Studies with hordein mRNA (Matthews, J. A., Coleman, A. and Miflin, B. J. unpublished results) have also shown that this message can be translated and the products processed to the correct size by oocytes. Further we have isolated RNA-free vesicles from barley endosperm RER and shown that they are capable of processing and protecting the hordein translation products when present, during translation of the hordein mRNA in a wheat germ system, in a manner strictly comparable with the dog pancreas system. These results should not be confused with the findings of the Carlsberg group (Cameron-Mills and Ingversen, 1978; Cameron-Mills et al., 1978) in which they claim that the processing can occur post-translationally, will occur with polysome run-off products, and involves products of intermediate sizes, none of which observations are consistent with the Blobel and Dobberstein hypothesis. Besides providing information on how storage proteins are synthesized and translocated the results demonstrate the homology of processing signals between plant and animal systems.

Legumes

In contrast to the cereal systems studied the legume work presents a more confused picture. Thus Sun et al., 1975, 1978; Beachy et al., 1978, Higgins and Spencer, 1977, were unable to demonstrate that the legume storage proteins were preferentially synthesized on membrane bound polysomes. Recently however Bollini and Chrispeels (1979), using techniques akin to the cereal workers, were able to demonstrate that this was the case for the 7S storage proteins of Phaseolus vulgaris and Püchel et al., 1979 obtained similar results for V. faba globulins.

Studies with isolated poly A^+ RNA fractions have shown that the proteins produced may be immunologically identical to the authentic storage proteins of P. sativum (Croy et al., 1980a,b; Higgins and Spencer, 1980). However, there is no clear agreement as to whether or not the products are of the same size, or larger than, the authentic polypeptides. Studies by Higgins and Spencer (1980) indicate that processing of the products of polysomal RNA, by membrane vesicles, leads to a reduction in size so that they approximate to authentic polypeptides. In contrast others (Croy et al., 1980a,b) claim that there is no difference between the migration of the products of pea cotyledon polysomes and of the poly A^+ RNA derived from them (although comparative, twin-track, gels are not presented in the papers). Translation of mRNA from soybean cotyledons leads to products identified as the α and α^1 subunits of the 7S (glycinin) storage proteins with electrophoretic mobilities similar to those of the authentic polypeptides (Beachy et al., 1980a,b). Studies with Phaseolus vulgaris mRNA specifying the G-1 polypeptides indicate that the major products migrate similarly to the β and γ subunits and a band equivalent to the α subunit is not seen. This is ascribed to the greater degree of glycosylation of the native subunit and it is suggested that the non-glycosylated α has a similar mobility to that of γ subunit (Hall et al., 1980). At present little is known as to what if any signal sequences may be present in the translation products of legume storage protein mRNA.

Work with isolated membrane systems from developing legume cotyledons has demonstrated the presence of N-acetylglucosamine and mannose transferases associated with the RER and there is also some association of mannosyl transferases with membranes which may contain the Golgi apparatus (Nagahashi and Beevers, 1978). Subsequent studies have shown that particulate extracts of P. vulgaris cotyledons incorporate N-acetylglucosamine into vicilin-like polypeptides (Davies and Delmer, 1980).

GENETICS

Chromosomal Location of Structural Genes

Although storage proteins are complex mixtures, they appear to be coded by a relatively small number of loci. Evidence for the location of these loci comes from several types of study: classical genetic analysis using crosses between different genotypes, the analysis of protein fractions from lines with an altered complement of chromosomes and, most recently, *in situ* hybridization of radioactive mRNA and cDNA with the loci. Most information is available for cereals and these will be considered first.

Barley. Genetic analysis of crosses between cultivars with different 'B' and 'C' hordein patterns shows that these groups are coded for by codominant alleles at separate but linked loci, called *Hor 2* and *Hor 1* respectively, located on either side of the *Mla* locus on chromosome 5. Although each locus codes for a number of polypeptides, 1-D analysis of hordein fractions from F2 seed failed to show any recombination within either locus (Oram *et al.*, 1975; Shewry *et al.*, 1978c; Doll and Brown, 1979; Jensen *et al.*, 1980). To confirm that this was not due to comigration of polypeptides on 1-D systems or to problems in scoring heterozygotes in the triploid endosperm, Shewry *et al.*, (1980c) made a detailed study using 2-D and high resolution 1-D SDS-PAGE analysis of F2 material and homozygous doubled monoploid lines produced by the *Hordeum bulbosum* technique (Jensen, 1976). Although the equivalent of a total of 347 gametes were analysed, no recombination within either locus was found. However, it is unlikely that recombination within a map distance of 0.010 centiMorgans would have been detected.

Wheat

Hexaploid bread wheat has three genomes, each with seven pairs of chromosomes, which are probably derived from related diploid grasses of the genera *Triticum* and *Aegilops*. The chromosomes contributed by these three genomes are homologus but non-pairing (homoeologous) and probably carry much shared genetic information. Because of this it is possible to produce lines in which a pair of chromosomes normally contributed by one of the genomes are absent, and are replaced by an additional pair from one of the other genomes. Shepherd (1968) and Wrigley and Shepherd (1973) have used such compensating nullisomic-tetrasomic lines of the cv. Chinese Spring to determine the location of gliadin genes. Using 1-D SGE (Shepherd, 1968) and 2-D IEF/SGE (Wrigley and Shepherd, 1973) they showed that almost all of the gliadins could be assigned to the homoeologous chromosomes of groups 1 and 6. More recently Kasarda et al. (1976) confirmed many of these assignments using 1-D aluminum lactate PAGE of some of the same lines and of substitution lines of Cheyenne in Chinese Spring. They also used 6A α ditelocentrics of Chinese Spring to locate the genes coding for the 'A' gliadin group of α gliadins on the α arm of chromosome 6A. From these studies it can be concluded that the α gliadins are coded mainly by 6A and 6B, the ω-gliadins mainly by 1B and 1D and the β and γ gliadins by chromosomes 1 and 6 of all three genomes.

More recently Brown and co-workers have used 2-D analysis of total endosperm proteins to analyse aneuploids and interspecific and intraspecific substitution lines of Chinese Spring. Although the differences in analytical procedures makes it difficult to make detailed comparisons between their results and those of previous workers, they also showed that gliadins are controlled only by the homoeologous chromosomes of groups 1 and 6 (Brown et al., 1979; Brown and Flavell pers. commun.). Their results also suggested that there could be regulatory genes on group 2 chromosomes which caused deletion of specific proteins whose presumed structural genes are on chromosome 1D and 6D and definitive proof is awaited (Brown and Flavell, pers. commun.).

Results from the analysis of aneuploid and substitution lines suggest that the high mol. wt. gliadins (termed glutelins by some) are also coded by group 1 chromosomes. Bietz et al. (1975) showed that two polypeptides each of Chinese Spring were coded by 1D and 1B chromosomes, and the assignment of 1D subunits has recently been confirmed by Brown and Flavell (1979 and pers. commun.). In other varieties there are high mol. wt. polypeptides coded by for the 1A chromosome (Payne et al., 1980). The inheritance of the patterns in crosses shows simple segregation with gene dosage effects.

This approach does not, however, tell us how the genes are arranged on the chromosomes. The presence in wheat, as in barley, of complex loci coding for groups of gliadin polypeptides is indicated by the genetic studies of Doekes (1973) and Mecham et al. (1978). Doekes (1973) showed that the gliadin pattern exhibited on starch gels could be divided into 6 or 7 sections which were inherited unaltered in F10 lines selected from intervarietal crosses. The α and γ gliadins appeared to constitute separate sections, the β gliadins, 2 sections and the ω gliadins 2 or 3 sections. A more detailed study was made by Mecham et al. (1978) who followed the inheritance of individual gliadin polypeptides in F1 and F2 seed of intervarietal crosses. They showed that, as in barley, the gliadin genes were codominant and the polypeptides appeared to be inherited as groups, for example the α 4-6 gliadins of Cheyenne and INIA 66R and the α_1, and α_2 gliadins of Justin were inherited as groups and were apparently allelic. Sosinov and colleagues (see Rybalka, 1980) have analysed a number of crosses and varieties and have identified a number of genetic loci Gld 1A, Gld 1B, Gld 1D, Gld 6A, Gld 6B, Gld 6C, which govern the patterns obtained on SGE and have also described allelic variants for several Russian varieties thus enabling the SGE patterns used in variety identification to be put on a rational basis.

Rye is a simple diploid which is closely related to wheat. However, Shepherd (1973) used substitution lines of rye chromosomes into Chinese Spring wheat to show that all the prolamin genes were located on a single chromosome which he called 1R, to emphasise its homology with chromosome 1 of wheat.

Maize

As in wheat, the prolamins of maize appear to be controlled by loci located on several chromosomes. The genes are again codominant and show simple Mendelian segregation in the F2 (Soave et al., 1978). Four IEF bands were shown by genetic analysis to be controlled by genetic elements 5.5 cM from the opaque-2 locus on the short arm of chromosome 7 (Soave et al., 1978, 1979). Although these were named Zp 1, Zp 2, Zp 3 and Zp 16, no recombination between them was observed and they are probably part of a complex locus equivalent to the Hor 1 or Hor 2 locus of barley. Two other structural loci coding for IEF bands (Zp 6 and Zp 12) segregated independent of each other, opaque-2 and Zp 1-3. The authors concluded that loci specifying zein storage proteins were located in at least three positions in the genome. Further structural loci on chromosomes 4 (long and short arms) and 10 (long arm) were located using TB-A translocation and trisomic stocks (Valentini et al. 1979; Soave et al., 1979). More recently Viotti et al. (1980) used in situ hybridization of ^{125}I-mRNA and ^{3}H-cDNA to locate zein structural genes on the long arms of chromosomes 4 and 5, the short arm of chromosome 7 and the distal segment of the long arm of chromosome 10. The two hybridization sites on chromosomes 7 and 10 are found near the opaque-2 and opaque-7 loci. The locations on chromosomes are consistent with the previous reports (Soave et al., 1978, 1979; Valentini et al., 1979) but this is the first evidence on zein genes on chromosome 5.

Legumes

Less is known about the genetic control of legume storage proteins. However the available evidence suggests that they are again coded for by codominant alleles, and that multiple structural loci also exist.

Hall and coworkers have shown that lines of Phaseolus differ in the subunit pattern of G1 globulin (Romero et al., 1975; Hall et al., 1977) as measured by SDS-PAGE. Whereas some lines had subunits of 53, 47 and 53 kd, in others the subunit mol. wts. were 50.5, 47 and 43 kd. Both 53 and 50.5 kd subunit types were present in F1 seed from crosses between the phenotypes, and in the F2 patterns segregated independently in ratios indicating control by a single locus.

In Pisum, Thomson and coworkers (Thomson and Schroeder, 1978; Thomson and Doll, 1979) have reported structural loci, designated Vca, Vcb, Vcc, Vce and Lga, Lgb, Lgc which control variations in the electrophoretic patterns, produced on SDS-PAGE and urea-PAGE, of the subunits of vicilin and legumin respectively. The control of the 40 kd and 20 kd subunits of legumin appears to be complex in that the products of more than locus migrate with the same molecular weight. Similar findings have been reported by Casey (1979b) using 2-D separation systems, where he also found complex segregation patterns in the F_2 for one sub-group of 40 kd polypeptides. Although a final conclusion cannot yet be drawn for the genetic control of the pea storage proteins the results so far suggest that considerable gene duplication may be involved.

Homology of Storage Polypeptides - Evidence for Duplication and Divergence of Structural Genes

The polymorphism and microheterogeneity of storage proteins has been referred to above. Chemical analysis of the polypeptides suggests that this is the result of gene duplication and divergence.

Cereals

Bietz, Huebner and Rothfus (1970) have shown similarities in the peptic peptide maps of purified α, β and γ gliadins, indicating some sequence homology between the three groups. A similar approach has been used to determine structural homology between hordein polypeptides. Holder and Ingversen (1978) reported similarities in the peptide maps of chymotryptic digests of ^{35}S-methionine-labelled 'B' and 'C' hordein polypeptides. However, the number of labelled peptides from the 'C' polypeptides was surprisingly large considering the low number of methionine residues present (only one or two residues/mole Schmitt, 1979; Shewry *et al.*, 1980d) and their probable location close to the C-terminus (Schmitt, 1979; Schmitt and Svendsen, 1980; Shewry and Kasarda, unpubl. results). The homology of these two groups of polypeptides has not been confirmed.

We have recently made a detailed study of the degree of polymorphism and homology of the 'B' polypeptides coded by the *Hor-2* locus (Faulks, Shewry and Miflin, manuscript in preparation Miflin *et al.*, 1980c). Comparison by 2-D analysis of hordein fractions from 8 cultivars showed that at least 47 different polypeptides could be recognised with each variety having between 8 and 17 polypeptides (excluding trace components). The distribution of polypeptides suggested that some recombination within the locus had taken place during the evolution of the varieties. The polypeptides could be divided into several groups on the basis of their cyanogen bromide peptide maps and it is suggested that in at least some cultivars these might be coded by structurally distinct parts of the locus.

Evidence for the structural and genetic relatedness of zein polypeptides comes from the recent studies of Park *et al.* (1980). They showed that the zein mRNAs of Illinois High Protein (IHP) maize could be classified into three groups on the basis of their binding to cloned cDNA copies of IHP mRNA. Three other lines had mRNAs related to those of IHP. The zein polypeptides synthesised by messages within a family had similar mol. wts., but varied in isoelectric point. It is not known if these families of mRNAs were the products of the same or separate structural loci.

More precise information on polypeptide homology is given by N-terminal sequence analysis (Table 5). A number of , and gliadins have been analysed (Kasarda *et al.*, 1974; Bietz *et al.*, 1977) showing the presence of two types of N-terminal sequence.

Table 5

N-terminal amino acid sequences of purified polypeptides of barley and wheat and mixtures of prolamins from rye and maize.

Cereal	Sample	1	2	3	4	5	6	7	8	9	10	11	12	13	14	15	16	17	18	19	20	21	22	23	24	25	Refs
Common (bread) Wheat (T. aestivum)	α2, α8, α9, α10, α11, α12	V	R	V	P	V	P	Q	L	Q	P	Q	N	P	S	Q	Q	Q	P	Q	E	Q	V	P	L	V	1, 2
(T. aestivum)	β5 gliadin	V	R	V	P	V	P	Q	L	Q	P	Q	N	P	–	Q	Q	Q	P	Q	E	Q	V				2
(T. aestivum)	γ1 gliadin	V	R	V	P	V	P	Q	L	Q	P	Q	N	P	<u>S</u>	Q	Q	Q	P	Q	Z	Q	V	P	L	V	2
(T. aestivum)	γ2 gliadin	N	I	G	V	D	P	W	G	Q	V	Q	W	L	P	Q	Q	<u>Q</u>	V	P	<u>Q</u>	L	Z	<u>Q</u>	<u>P</u>		2
(T. aestivum)	γ3 gliadin	N	M	G	V	D	P	W	G	Q	V	Q	W	V	P	Q	<u>Q</u>	L	Q		<u>Q</u>	<u>Q</u>	<u>Q</u>			Z	2
Diploid wheat (T. monococcum)	ω -gliadin	R	Q	L	N	P	S	D	Q	E	L	Q	S	P	Q	Q	L	Y	P	Q	Q	P	Y	P	Q	Q	3
Barley	'C' polypeptide	R	Q	L	N	P	S	S	Q	E	L	Q	S	P	Q	Q	S	Y	L	Q	Q	D	Y	P	Q	N	3
Rye	Water-soluble secalin	Primary sequence																									
		N	M	Q	V	P	P	S	G																		
		Secondary residues																									
		A	Q	L	N	G	S	Q	Q																		
		L				N																					
Maize	Total zein	Primary sequence																									
		T	I	F	P	Q	C	S	Q	A	P	I	A	I	L	L	Q	P	Y	L	P	V	A	V	M	G	5
		Secondary residues																									
			F		I				L									P		F				I	A	F	
Maize	22kd band	S/F*	I	I	P	Q	S	S	L	A	P	S	A	I/S*	I												6

Primary sequences only are shown for the purified polypeptides. Secondary residues were present at some positions in the , 2 and 3 gliadins. Residues underlined are subject to uncertainty. High resolution polyacrylamide gel electrophoresis showed that 2, 8-11 and 1-3 were heterogeneous.
* A 1:1 ratio of two residues was observed. Standard 1 letter amino acid abbreviations have been used.

References
1. Kasarda et al. (1974).
2. Bietz et al. (1977).
3. Shewry et al. (1980a). Similar sequences have been reported by Bietz (1980), Shewry et al (1980b) and Schmidt and Svendsen (1980).
4. Autran et al. (1979).
5. Bietz et al. (1979)
6. Larkins et al. (1980)

The α_2, α8-12, β 1 and γ 1 gliadins had identical primary sequences for over 20 amino acids, although a number of the preparations were heterogeneous resulting in the presence of secondary amino acid residues at some positions. The γ 2 and γ 3 gliadins had a distinctly different type of sequence, and also had a number of secondary residues and differed from each other at several positions. Comparison of these two types of sequence, called the α and γ types by Autran et al. (1979), shows some homology, for example proline is always present at position 6 and glutamine at positions 15 and 16. As far as is known at present the α type gliadins are coded by the homoelogous chromosomes of group 6 and the γ type by group 1. A third type of gliadin sequence, which shows no homology with the α or γ types, was recently reported for a purified ω-gliadin component from Triticum monococcum a diploid wheat which is probably related to the progenitor of the A genome of hexaploid bread wheats (Shewry et al., 1980a). The ω-gliadins are only present as a small proportion of the total gliadin fraction, and sequence analysis of a total gliadin preparation showed the presence of only the α and γ -type sequences (Autran et al., 1979).

Purified groups and individual B and C polypeptides from barley have also been analysed. The B hordein polypeptides are very similar to α , β and γ gliadins in molecular weight and amino acid composition, but are N-terminally blocked (Shewry et al., 1980d,e; Schmidt and Svendsen, 1980). The 'C' polypeptides are similar to ω-gliadins in many of their properties, and the N-terminal sequence of a purified polypeptide was found to be similar to that of the ω-gliadin from T. monococcum (Shewry et al., 1980a). Similar sequences have been demonstrated for three purified 'C' polypeptides from a different cultivar (Shewry, Lew and Kasarda, unpublished results), 'C' polypeptide mixtures (Bietz, 1980; Shewry et al., 1980e) and a partially pure 'C' polypeptide (Schmitt and Svendsen, 1980). This suggests a high degree of homology between the polypeptides coded by the locus.

N-terminal sequence analysis of a whole prolamin mixture has also been used by Bietz et al. (1979) to determine the degree of homology of zein polypeptides. The major sequence was unrelated to those found in other cereal prolamins, and secondary residues were detected at only 5 positions out of the first 20, 11 positions of the first 33. The authors concluded that this indicated a high degree of homology of the polypeptides in the fraction. A 14-residue N-terminal sequence which differs in several positions has been reported for a group of zein polypeptides of mol. wt. approx. 22 kd by Larkins et al. (1979).

However both total zein and the 22 kd band consist of a number of polypeptides (Miflin and Shewry, 1979b, Fig. 1) and since the analysis of prolamin mixtures would not detect quantitatively minor sequences or N-terminally blocked polypeptides, it is probable that the range of variability in zein sequences is greater than those reports suggest.

Wheat, barley and rye are closely related members of the Triticanae, and probably evolved from a common ancestral stock. Smith and Flavell (1974) have concluded on the basis of DNA hybridization data that the divergence of wheat and rye occurred after their separation from barley. However, whereas in rye and barley all the prolamins are coded for by genes on a single chromosome, in wheat two chromosomes (1 and 6) from each of the contributory genomes are involved (Wrigley and Shepherd, 1973). Shepherd (1973) has suggested that originally these genes were located on one chromosome, and that translocation of part of this to another chromosome gave rise to the common ancestor, which, in turn, differentiated into the progenitors of the common wheat genomes. We can speculate that the common ancestor of wheat, barley and rye had two storage protein loci which were located on the same chromosome and coded for ancestral forms of the ω gliadin/C hordein and γ-gliadin/B hordein types of polypeptide. In barley this chromosome became chromosome 5 and further duplication and divergence of the genes at each locus resulted in the present day Hor-1 and Hor-2 loci. A similar process of duplication and divergence may have taken place on chromosome IR of rye. In the progenitor of wheat, however, the γ locus may have duplicated and part of it translocated to chromosome 6 where it diverged to give the α-type sequence. Further duplication and divergence at all three loci then resulted in the wide spectrum of gliadin polypeptides found in modern wheats.

Legumes

Although polymorphism of the legume storage protein exists (see above) less is known about the homology of the polypeptides.

Tryptic digests of acidic subunits A1-3 of glycinin show a considerable degree of homology between the three and also a limited similarity to the A_4 (serologically distinct) subunits. More recently Moreira et al. (1979) have purified and characterized six acidic and four basic subunits of glycinin. N-terminal amino acid sequences showed considerable homology between the members of the individual families of acidic and basic subunits, suggesting that the members of each family arose from a common ancestral gene.

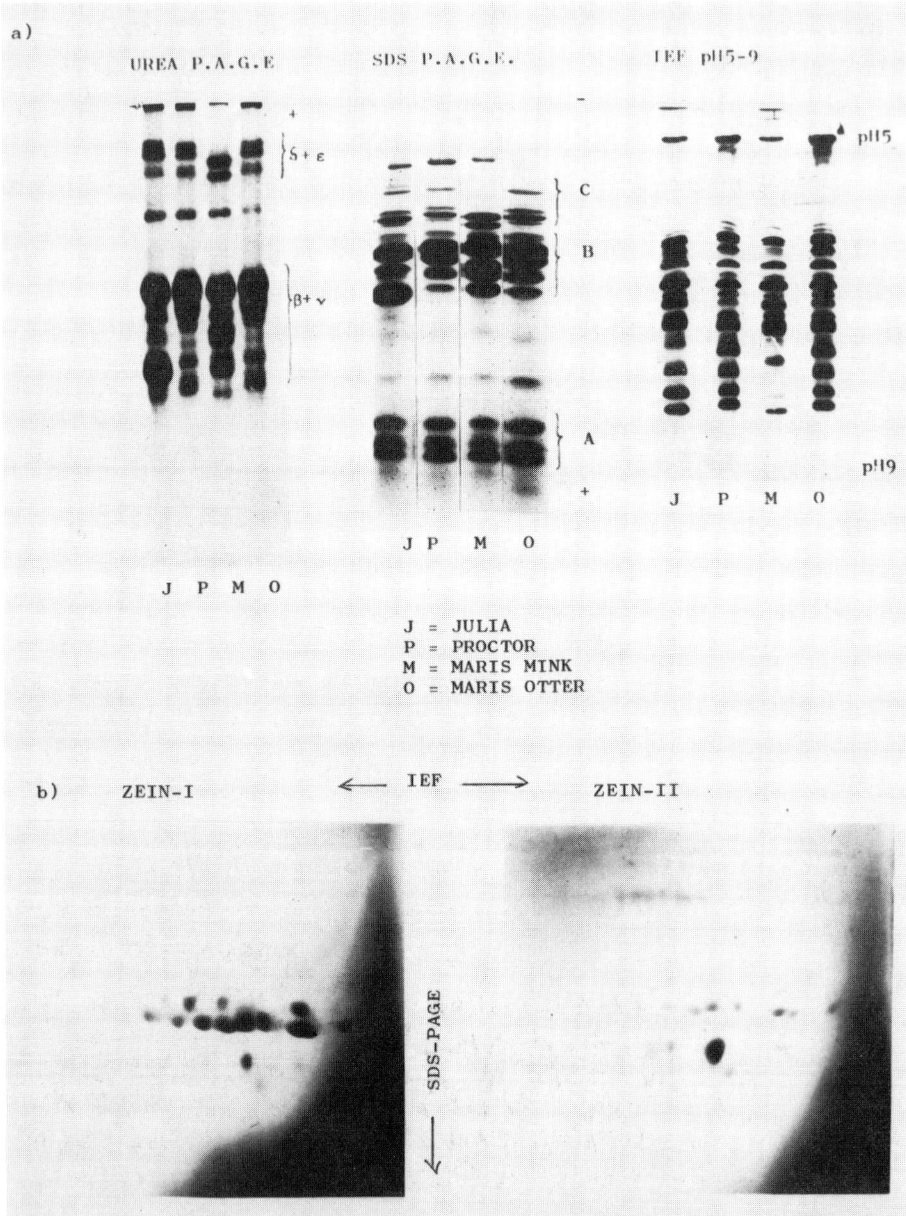

Figure 1 (a) Hordein separated by different techniques (b) Zein separated by 2-D IEF/SDS-PAGE. Zein-I extracted in the absence and subsequently Zein-II extracted in the presence of mercaptoethanol.

Ma and coworkers (A. Ma, A. Bliss, and T. C. Hall, pers. commun.) have recently made a detailed study of the homology of three polypeptide subunits of G1 storage protein from Phaseolus vulgaris. The polypeptides were subjected to limited proteolysis with five different proteolytic enzymes (trypsin, chymotrypsin, papain, proteinase K, protease V8) and to cleavage with CNBr and BNPS-Skatole, and the resulting peptides separated by SDS-PAGE. Comparison of the peptide maps of the three subunits indicated at least 66% homology in amino acid sequence.

Even more widely related proteins appear to be closely homologous thus Gilroy et al., 1979 have recently compared N terminal amino acid sequences (29 residues) of the (total mixed) basic subunits of glycinin from soybean with those of vicilin from V. faba (which are not immunologically cross-rective, Dudman and Millerd, 1975) and found evidence for the presence of the same residue at 27 positions; although more than one residue was found at certain positions indicating the microheterogeneity of the subunits. The sequences found were similar to those subsequently reported for soybean purified 11S-basic subunits (Moreira et al., 1979).

Regulatory Genes which affect Protein Synthesis and Quality

Cereals

In 1964 Mertz and co-workers described the identification of a regulatory mutant gene, called opaque-2, which caused an increase in the lysine content and hence nutritional quality of maize grain (Mertz et al., 1964)(Table 5). Subsequently a number of other mutant genes were identified, floury-2 (Nelson et al., 1965), opaque-7 (McWhirter, 1971), opaque-6 and floury-3 (Ma and Nelson, 1975), brittle-1, brittle-2, shrunken-1, shrunken-2, shrunken-4, sugary-1 (Misra et al., 1975) and De*40 (Salamini et al., 1979). All are spontaneous mutations identified by mutant endosperm phenotypes caused by changes in carbohydrate metabolism - hence the designations as floury, opaque etc. This disruption in carbohydrate metabolism results in decreased seed weight (from 40-90% of normal lines, Table 6) and yield. However it is possible to modify these effects by using different genetic backgrounds (Nelson, 1979) or modified genes (Vasal, 1975). The increase in lysine varies from over 100% see Table 6. The only dominant mutant gene is De*30 (Salamini et al., 1979) but floury-2 and floury-3 exhibit semidominance (Ma and Nelson, 1975; Jones, 1978).

Table 6

The chromosomal location and characteristics of high lysine mutant genes of Maize and Sorghum

Mutant gene	Chromosome	Lysine (% of control line)	% recovered N: Prolamin	Glutelin	Salt-soluble protein	NPN	Endosperm wt (% control)	Ref
Maize								
Normal(Oh43)	-	-	64.8	26.5	5.8		-	1
Opaque-2	7	218	35.3	43.2	13.6		68	
Floury-2	4	169	58.1	29.6	9.2		87	
Sugary-1	4	112	49.0	31.9	11.9		73	
Shrunken 1	9	119	56.0	30.7	8.2		71	
Shrunken 2	3	169	38.8	38.6	12.3		43	
Shrunken 4	5	187	38.5	31.9	25.7		27	
Brittle 1	5	144	52.4	35.7	8.8		82	
Brittle 2	4	206	41.5	36.6	12.1		41	
Normal(W22)	-	-	55.9	33.8	6.9		-	2
Opaque-7	10	165	32.3	48.3	16.6		60	
Normal	-	-	62.4	29.6	5.6	0.4	-	3
Opaque 6	?	214	28.8	37.7	20.6	8.1	84	
Normal	-	-	55.3	32.7	6.5	1.4	-	3
Floury 3	8?	168	41.1	41.2	14.4	4.7	44	
Normal	-	-	55.4	38.6	6.0		-	4
De*30	?	176	38.8	46.8	12.4		63*	
Sorghum								
Normal(P721)	-	-	50.2	40.8	9.0		-	5
P721 opaque	?	147.5	25.2	46.2	28.6		-	
IS 11167	?	160	29.7	47.2	23.1		-	

With the exception of IS 11167 comparisons are made with isogenic or near-isogenic normal lines.
NPN = non-protein nitrogen (free amino acids in Ref. 3 only).
* whole seed weight.

References

1. Misra et al, 1975 (Cereal Chem 52, 161). Endosperm wt only from Tsai et al, 1978 (Biochem Genet. 16. 883). Zein is ZI + ZII.
2. Misra et al, 1975 (Cereal Chem 52, 161). Endosperm wt only from Dalby & Tsai 1975 (Crop Sci. 15513). Zein is ZI + ZII
3. Recalculated from Ma & Nelson 1975 (Cereal Chem. 52-412). Zein is ZI only.
4. Salamini et al, 1979 (AEA book). Zein is ZI + ZII. Glutelin fraction includes residue.
5. Guiragossian et al, 1978 (J Ag Fd Chem 26. 219). Prolamin is prolamin I + II. IS 11167 is an Ethiopian line with the hl gene.

The identification of high-lysine genes in maize stimulated the search for similar mutants in other cereals. The first high lysine gene discovered in barley, designated lys, was in the Ethiopian line Hiproly (Munck et al., 1970). Although Hiproly itself is very primitive with small shrunken seed and low yield, several breeders have successfully incorporated the gene into lines with plump well-filled grain and yields only slightly below normal (Berdahl and Bhatty, 1977; Persson and Karlsson, 1977)(Table 7). Other high lysine lines have been produced by mutation breeding (Table 7). All appear to be single mutant genes which, with the exception of Riso 8 (lys 4d) are recessive. Although the lys (Hiproly) and lys 3a (Riso 1508) genes are both located on chromosome 7 they are unlinked (Karlsson, 1977). The notch lys 95 and lys 449 genes have not been mapped, but the notch genes are not allelic to each other or to lys (Bansal and Bhaskaran, 1973). However some high lysine genes are allelic, lys 3a (Riso 1508) with lys 36 (Riso 18) and lys 3c (Riso 19), and lys 5f (Riso 13) with lys 6g (Riso 29) and lys 5h (Riso 86). The increased lysine content (to + 36%) is generally less than in maize, but except for Riso 7 is accompanied by decreased seed size and yield. In Riso 13, 29, 1508 and 86 this has been shown to be due to pleiotropic effects of the high lysine genes (Koie and Kreis, 1978; Kreis, 1978; Koie et al., 1976; Koie and Doll, 1979). Consequently although the lys 3a gene has been used intensively for breeding, no commercially valuable, high-yielding, high-lysine line has resulted. Even though Riso 7 has a relatively small increase in lysine (+ 11%) this is not associated by decreased seed size and it may have more potential for breeding.

Two high lysine genes in sorghum were discovered by screening for mutant endosperm phenotypes (Table 6). The Ethiopian high lysine gene (hl) results in a floury endosperm type with 15-17% protein (compared to 12% in control lines) and up to 80% more lysine (Singh and Axtell, 1973). This is, however, accompanied by a yield reduction in all backgrounds (Axtell et al., 1979). The P721 opaque gene was found by screening mutagen-treated seed (Axtell et al., 1979). It is a single partially dominant gene which results in up to 60% more lysine accompanied by a 11-14% decrease in seed weight. In some backgrounds, however, a compensatory increase in seed number results in high yields.

Table 7 The origin, characteristics and chromosomal location of high lysine mutants of barley

Gene	Origin	Parent line	Mutagen	Chromosomal location	Dominant or Recessive	Yield (% of parent line)	Lysine content (% of parent line)	1000 seed wt (% of parent line)	Ref
Lys	Hiproly	Ethiopian		7	recessive	95-99*	123-125*	-	1
	Risø mutant								
	29	Carlsberg II	EMS	not known		85	109	88	2
	86	Carlsberg II	EMS	not known		79	105	79	2
	Risø mutant								
	7	Bomi	nth			89	111	100	2
	8	Bomi	EMS	5	dominant or semi-dominant	79	109	84	2
	9	Bomi	EMS	not known		54	103	65	2
	13	Bomi	EMS	6	recessive	72	123	79	2
	16	Bomi	nf	1	recessive	84	108	86	2
	17	Bomi	nf	1	recessive	57	110	85	2
	527	Bomi	rays	6	recessive	69	112	80	2
Lys 39	1508	Bomi	EI	7	recessive	77	136	88	2
	Notch 1	NP 113	EMS	not known	recessive	-	118	84	3
	2	NP 113	EMS	not known	recessive	98	117	81	3
Lys 95	Lys 95	Perga	EMS	not known	recessive	-	135	72	4
Lys 449	Lys 449	Perga	EMS	not known	recessive	77	130	73	4

* based on comparison of the high lysine Hiproly-derived lines Sv 74865 and Sv 75235 with the control variety Ingrid (Persson & Karlsson, 1977).

References
1. Karlsson, 1976; Persson & Karlsson, 1977; Munck et al, 1970.
2. Doll et al, 1974; Doll, 1976; Karlsson, 1977; Jensen, 1979a, b; Doll, 1980.
3. Balarāvi et al, 1976; Bansal et al, 1977.
4. DiFonzo & Stanca, 1977; DiFonzo. Pers. Commun.

EMS, ethyl methane sulphonate; EI, ethylinimine; nth, thermal neutrons; nf, fast neutrons.

The biochemical effects of high lysine genes have been extensively studied, and detailed accounts have been given in recent reviews by Mertz (1976); Nelson (1979; 1980), and Miflin and Shewry (1979a). All the mutants studied appear to be low prolamin mutants in that the total and relative amount of this lysine-poor fraction is reduced (Table 8). The decrease in prolamins is compensated for by increases in other more lysine-rich fractions. In most of the mutants there is an increase in non-protein amino acids and in salt-soluble proteins and glutelins, although the mutants differ in the relative effects. With the possible exception of the lys gene of barley, there is no change in the lysine content of the non-prolamin fractions rather the increased lysine content of the seed appears to be entirely a result of the change in balance of the protein fractions. Although it as originally thought that the main effect of the lys gene was to increase the amount of certain lysine-rich proteins in the salt-soluble fraction (Munck, 1972; Inversen and Koie, 1973) recent analyses indicate that a decrease in prolamins may also contribute to the increase in total grain lysine (Miflin and Shewry, 1979a).

Some high lysine genes also cause alterations in the polypeptide composition of the prolamins. Comparison of hordein fractions from Riso 1508 and its parent line (Bomi) shows that all the major 'C' polypeptides and a number of 'B' polypeptides are absent, and that multiple spots are present in the positions of other 'B' polypeptides (Shewry et al., 1980b Fig. 2). Several other Riso mutants have changes in the relative amounts of 'B' hordein polypeptides only (Koie and Doll, 1979), while the lys 95, lys 449 and notch mutations have only minor effects or no effect at all on the hordein composition (Shewry et al., 1980b). Similarly, several of the mutant genes of maize have differential effects on the zein polypeptide composition (Lee et al., 1976; Tsai et al., 1978; Jones, 1978; Soave et al., 1976, 1979; Gianazza et al., 1976; Di Fonzo et al., 1977, 1979). For example, whereas the floury-2 gene seems to affect both the major SDS-PAGE bands (the Z19 and Z23 components of Soave et al., 1979) equally, the opaque-2 and opaque-7 genes preferentially decrease the Z23 and Z19 components respectively. In contrast the De*30 gene appears to affect only the Z23 component without affecting the Z19 (Soave et al., 1979). The opaque-2 gene causes a greater inhibition of the most alkaline zein bands separated by IEF.

Table 8

The protein fractions of High Lysine Barley Lines

	% TOTAL N							
	Total seed N (%)	Lysine (%)	NPN	Hordein	Salt-Soluble Protein	Glutelin	Residue	Ref
Hiproly	3.64	3.58	8.3	41.1	19.4	25.6	4.5	
Mona	2.92	2.84	9.0	52.5	13.5	15.1	3.6	
Birgitta	3.12	2.81	8.3	55.9	14.5	15.6	3.1	1
Sv 73608*	3.49	3.48	7.0	44.6	18.1	25.5	3.5	
Sv 76457*	3.53	3.42	7.6	40.7	21.5	22.6	3.7	
NP 113	1.80	2.9	6.3	47.3	13.3	24.3	3.0	
Notch-1	2.03	3.9	9.3	32.0	15.8	33.8	3.6	2
Notch-2	1.93	3.9	11.1	29.0	17.1	32.8	3.5	
Perga	1.71	3.2	8.1	45.8	10.9	22.5	4.3	
Lys 95	2.07	4.1	13.2	31.2	11.5	35.4	6.4	2
Lys 449	2.29	4.2	12.1	35.2	11.1	33.6	6.2	
Bomi	1.47	3.3	11.6	45.2	15.6	18.0	5.2	2
Risø 1508	1.63	4.9	19.9	15.6	25.4	26.5	6.3	
Carlsberg II	1.37	3.5	5	40		55		
Risø 29	1.54	4.3	5	37		58		3
Risø 86	1.60	4.4	5	38		57		
Bomi	1.36	3.5	5	41		55		
Risø 527	1.56	4.2	8	31		62		
Risø 1508	1.49	5.1	8	15		77		3
Risø 7	1.39	3.8	6	33		61		
Risø 8	1.53	4.3	7	29		64		
Risø 13	1.69	5.0	10	27		62		
Bomi	1.45	3.5	6	44		50		
Risø 16	1.75	4.6	5	38		58		3
Risø 17	1.85	5.1	6	37		57		

NPN = non-protein nitrogen

*Sv 73608 = Mona + Hiproly. Sv 76457 = Sv 73608 + Birgitta

Refs. 1. Miflin & Shewry, 1979a. 2. Shewry et al 1980b. 3. Køie & Doll, 1979

Because the high lysine genes of maize and sorghum were selected on the basis of mutant endosperm phenotype it is not surprising that in all cases there are defects in carbohydrate metabolism which result in decreased seed weight. However, similar decreases in seed weight are also present in most of the barley mutants which were selected by DBC. This problem is discussed further below.

Legumes

Recently Davies (1980) has discovered mutations in P. sativum that may have somewhat analogous effects, to those reported for maize and barley, on storage protein fractions. Thus the presence of the wrinkled mutation specified by the r_a locus leads to a reduction of the proportion of legumin in the seeds. Analysis of crosses shows that the structural genes for the 40 kd subunits are about 7 recombination units away from r_a.

Structural Mutations at Storage Protein Loci

Although a number of mutations at 'regulatory' loci are known to affect storage protein synthesis in cereals, only one possible structural mutation has been reported. This may be because most mutation breeders have used chemical mutagens which cause single base changes in the DNA. Even if these changes resulted in amino acid substitution in the storage polypeptides, the mutations would probably remain undetected because of the complex and multigenic nature of the fraction and the difficulty of separating polypeptides differing in only a single amino acid.

Riso mutant 56 is a high lysine line of barley which was produced by γ-ray treatment which can cause chromosome breakage of the variety Carlsberg II (Doll et al., 1974). The high lysine trait is caused by a mutation which maps either at, or close to, the Hor-2 locus (Doll, 1980) on chromosome 5. Although it could be at a closely linked regulatory gene, analysis of the hordein fraction suggests that it is a structural mutation which results in deletion of part of the Hor-2 locus. The total amount of hordein is decreased (which results in the high lysine character) due to a decrease in 'B' polypeptides (coded by the Hor-2 locus) only. There is also a change in the relative amounts of the different 'B' polypeptides (Fig 2, Shewry et al., 1980b), although most of them are present in decreased amounts the two of lowest mol. wt. (arrowed in Fig. 2) form a higher proportion of the total. The 'C' polypeptides do not change in pattern and probably increase in total amount to partially compensate for the decrease in 'B' hordein. The increase in lysine is relatively small (≃ 13%) and there are also decreases in seed size (≃ 12%) and yield (≃ 23%)(Doll et al., 1974). It is not known, however, if all the effects are caused by the high-lysine mutation.

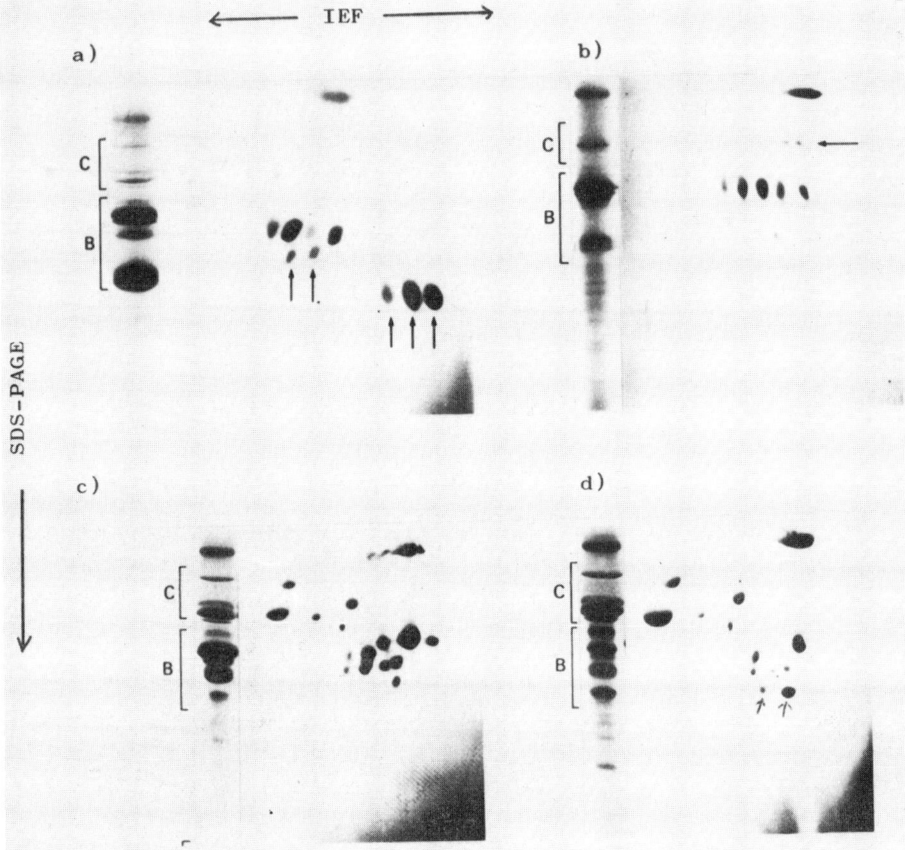

Figure 2. Two dimensional IEF/SDS-PAGE of hordeins from (a) Bomi (b) 1508 (c) Carlsberg II and (d) Risø 56. The arrows highlight changes between the mutants and the wild type (see also the text).

CONCLUDING DISCUSSION AND THE WAY AHEAD

As a result of the recent upsurge in interest a clearer picture of the nature of seed proteins is becoming apparent. Firstly, they are highly polymorphic and specified by a series of genes, probably derived by gene duplication, which have subsequently undergone mutation. The genes are clustered in complex loci residing in relatively few places on the chromosomes. The degree of duplication may have been selected for under conditions in which larger seeds were favoured (e.g. during domestication) as it probably affects the amount of storage polypeptides synthesized. The storage proteins of most of the cereals and the legumes are sharply different in their physical nature, amino acid composition, and means whereby they are deposited within the seed. However they have certain points in common; in both cases they have fractions that are extremely low in S, which probably provides an advantage in that plants are able to store surplus N under conditions in which S is limiting. This appears to be borne out by studies on the effect of N and S fertilization on the relative composition of different storage proteins and the amino acid composition of seeds (Blagrove et al., 1976; Davies, 1976; Byers et al., 1977).

From a consideration of seed physiology and from the effect of storage-protein-deficient mutations it is possible that the proteins play an important role in the efficiency of the seed development. If, as seems likely, phloem unloading depends on an osmotic concentration gradient across the phloem/storage compartment interface, then this will be increased by any process that sequesters osmotically active small molecules out of solution; conversely, the failure of the storage system to deal with monomers will decrease the rate of phloem unloading and hence the rate of transfer of metabolites into the seed. Because the osmotic pressure in the seed is largely dependent on the sum of salts, sugars and amino acids, increases in the level of any one of these will probably affect the rate of accumulation of the others within the storage compartment. Consequently it is not surprising that, when a mutation occurs that prevents the formation of one polymer, there is a concurrent decrease in the synthesis of the others and the overall seed size is affected. Furthermore, when double mutations affecting both protein and starch synthesis are combined, it would be expected (as is found) that there is a synergistic depression of storage function. Similarly links between ion accumulation and mutations in storage function might be predicted and in fact occur.

In conclusion, our review of storage proteins leads us to the opinion that the storage proteins perform an important physiological function which affects the ability of the seed to store total dry matter. In this way they directly contribute to crop productivity over and above their individual contribution to dry matter. Also, because of their amino acid composition they also largely limit the nutritional quality of the crop. The questions are what may be done in the future to improve the storage protein aspects of crop productivity?

Improvements in protein quantity

In general terms there is not a great priority in increasing the protein contents (i.e. % protein) of seeds except for specific purposes, such as wheat for breadmaking, since, for all but the youngest animals, there is probably a reasonable balance of protein and energy given that the protein were of the desired quality. Furthermore, because the protein content of the seed is greatly affected by the environment in which the plant is grown (particularly N fertilizer) and probably largely reflects the ratio of C: N supplied via the phloem from the vegetative part of the plant, there is a limit to what can be done by plant breeders. Nevertheless work in various centres, particularly with wheat, has shown that there are genes which are important in influencing seed protein content(see Johnson et al., 1979; Law et al. 1978). Whether or not the genetic capacity to store protein limits yields in normal varieties is not clear. It seems unlikely that this is the case in modern cereals where the total harvest and nitrogen harvest indices are very high, indicating very efficient storing of the nutrients present in the vegetative part of the plant. The same may not be true in legumes where the harvest indices are generally lower and an important proportion of the vegetative dry matter and nitrogen fail to be stored in the seed.

Improvements in protein quality

Four possible ways occur to us by which the amino acid composition of seeds might be improved.

Changes in the total amount of storage proteins Mutations which achieve this have been isolated in several cereals - they appear, by virtue of the location of the mutant genes in different chromosomes, to be of many different types, however almost all of those that fall into the 'regulatory' class cause smaller seeds. Unless this can be compensated for by increasing seed number it leads in turn to yield depression. Because of the nature of the system this may be inevitable and even though the yield of certain high lysine lines can be improved by plant breeding the rate of increase in yield of normal lines appears to be faster.

Changes in the relative proportion of the individual components Changing the relative proportion of B to C hordeins would lead, because of the slightly greater lysine content of the former (Table 4), to overall increases in lysine. Similarly changing the ratio of legumin to vicilin would increase the total content of S-amino acids. However this approach has several difficulties: because of the polymorphic nature of the protein and the complexity of the loci it might be difficult to achieve the required deletions of structural genes (alternatively, there may be regulatory genes that differentially affect different components but this has not so far been convincing demonstrated); also the degree of useful variation within the components of the storage proteins (certainly with respect to lysine in cereals) is not sufficiently large for major differences to be accomplished; and, finally, uncompensated depletions in part of storage protein complex may have yield penalties as discussed above. As yet, apart from our knowledge of the structural loci, we have too little genetic information on regulation of the synthesis of different components of the storage proteins to make many predictions.

Changes in the amino acid composition of the storage protein This approach has the advantage that if desired changes in composition could be introduced it may well have no effect on the storage function of the protein and thus yield. Unfortunately, because it is now apparent that the storage proteins are highly polymorphic and controlled by complex multiple genes, this is not likely to be readily achieved by simple mutation and plant breeding. For example to add two lysine residues/mole of zein would require up to 240 individual mutational events in which say, an Asn (AAU) was changed to Lys (AAA). An alternative approach is now becoming conceivable via genetic manipulation. This is envisaged as involving the following steps:- (a) isolation of the structural genes (or some of them); (b) chemical modification to increase the number of lysine codons; (c) probably, the synthesis of a series of linear duplications of the modified structural gene(s); (d) reintroduction of the genes into the plant, probably using recipients in which deletion of the structural genes had been induced. Such an approach is obviously futuristic but is nevertheless within the bounds of possibility since work on the isolation of genes, their chemical modification in vitro and their reintroduction in vivo is proceeding apace in animal systems (e.g. see Ooynen et al., 1979). We are engaged in such an approach in our laboratory and the results so far achieved by all workers in the field are encouraging. Thus there are now clones of double stranded cDNA for zeins (Park et al., 1980; Burr and Burr, 1980; Weinand et al., 1979) for hordeins (Brandt, 979, unpublished work from our laboratory) and for P. vulgaris G-1 proteins (Hall et al., 1980). Recently Hall and coworkers have identified and sequenced genomic clones for P. vulgaris G-1 proteins (T. Hall, personal commun.). Interestingly, they find that plant genes, like those from animals, have intervening sequences. Already one promising potential deletion mutant in barley exists (Riso mutant 56) and searches for more are in progress. The major difficulty is in obtaining recipient systems (e.g. regenerating protoplasts) for cereals and legumes although in this field there are now major efforts being made in many centres in different parts of the world (including our department at Rothamsted) and some mildly optimistic results are being obtained (Vasil and Vasil, 1980,). Independent of whether this approach is itself successful it will greatly increase our knowledge of the system and of the organisation of plant genes.

Changes in the nature of other nitrogenous fractions in the seeds Two possibilities occur one to increase a lysine-rich protein fraction and, two, to change the composition of the soluble pool of amino acids. Some evidence for lysine-rich proteins has been found in Hiproly (Ingversen and Koie, 1973) and it may not be long before such proteins are defined and their genetics determined. Providing they do not have a function that would be upset by increasing their proportion it might prove profitable to manipulate them.

The amount of certain amino acids (e.g. lysine, methionine and threonine) in the soluble pool is probably a function of the regulatory mechanism governing their synthesis. In a similar symposium in 1974 one of us proposed (Miflin 1976) that by selecting for mutants which have lost such regulatory properties it might prove possible to improve the amino acid composition of the seed. To do this we have grown hand-dissected embryos from M_2 seed on either lysine plus threonine (LT) or S-aminoethylcysteine (AEC) and selected those that grow. The toxicity of these compounds can be overcome by methione or lysine respectively (Bright et al., 1978) and so overproduction of these amino acids may be one way that plants can become resistant. So far we have recognized and partially characterized one AEC and three LT resistant mutants. One of the LT mutants (R2501) is capable of overproducing both threonine and methionine and has alterations in its enzymic control mechanisms (Bright et al., 1980 and unpublished results). This capability extends to the seed and there is a large increase in the free threonine pool such that there is a statistically significant increase in the total seed threonine of 10%. This increase exactly matches the extent of the deficiency of threonine in barley as measured against the standard FAO/WHO requirements. Although the gene is at present in mutated barley of unlikely agronomic use and has not so far been obtained in the homozygous state, the results demonstrate the potential use of the method.

Improvement of the crop quality of cereals in a manner not having deleterious effects on yield has not yet been achieved. Nevertheless there are a number of different approaches at present being explored and it should not be forgotten that high-lysine cereals do exist which can be used in circumstances where a small (10-20%) yield penalty can be tolerated.

ACKNOWLEDGEMENTS

We would like to acknowledge the help we have received from many colleagues in our work cited here - particularly A. Faulks, H. Pratt, S. Parmar, S. Burgess, J. Matthews, J. March, and D. Kasarda and the many stimulating discussions we have had with people both inside and outside our laboratory. We would also like to record our gratitude to Susan Wilson for all her efforts in producing this manuscript.

Alexandrescu, V., Chihaia, C., and Cosmin, O. (1976). Zein specificity in inbred lines of maize. Rev. Roum. Biochim. 13, 169-172.

Altschul, A. M., Yatsu, L. Y., Ory, R. L., and Engleman, E. M. (1966). Seed proteins. Ann. Rev. Plant Physiol. 17, 113-136.

Ashton, F. (1976) Mobilization of storage proteins in seeds. Ann. Rev. Plant Physiol. 27, 95-117.

Autran, J-C., Lew, E. J-L., Nimmo, C. C. and Kasarda, D. D. (1979). N-terminal amino acid sequencing of prolamins from wheat and related species. Nature 282, 527-529.

Axtell, J. D., Van Scoyoc, S. W., Christensen, P. J. and Efeta, G. (1979). Current status of protein quality improvement in grain sorghum. In: Seed Protein Improvement in Cereals and Grain Legumes Vol. II, IAEA: Vienna, pp.357-365.

Badley, R. A., Atkinson, D., Hauser, H., Oldani, D., Green, J. P., and Stubbs, J. M. (1975). The structure, physical and chemical properties of the soybean protein glycinin. Biochim. Biophys. Acta, 412, 214-228.

Bailey, C. J. and Boulter, D. (1970). The structure of legumin, a storage protein of broad bean (Vicia faba) seed. Eur. J. Biochem. 17, 460-466.

Bain, J. M. and Mercer, F. V. (1966). Subcellular organization of the developing cotyledons of Pisum sativum L. Aust. J. Biol. Sci. 19, 49-67.

Balaravi, S. P., Bansal, H. C., Eggum, B. O. and Bhaskaram, S. (1976). Characterization of induced high protein and high lysine mutants in barley. J. Sci. Fd. Agric. 27, 545-552.

Bansal, H. C. and Bhaskaran, S. (1973). Test for allelism between Hiproly and Notch mutants. Barley Genetics Newsletter, 3, 10-11.

Bansal, H. C., Srivastava, K. N., Eggum, B. O. and Mehta, S. L. (1977). Nutritional evaluation of high protein genotypes of barley. J. Sci. Fd. Agric. 28, 157-160.

Basha, S. M. M. and Beevers, L. (1976) Glycoprotein metabolism in the cotyledons of Pisum sativum during development and germination. Plant Physiology, 57, 93-97.

Beachy, R. N., Thompson, J. F., and Madison, J. T. (1978). Isolation of polyribosomes and messenger RNA active in in vitro synthesis of soybean seed proteins. Plant Physiology 61, 139-144.

Beachy, R. N., Barton, K. A., Thompson, J. F., and Madison, J. T. (1980a). In vitro synthesis of the and ^{1} subunits of the 7S storage proteins (conglycinin) of soybean seeds. Plant Physiol. 65, 990-994.

Beachy, R. N., Barton, K. A., Madison, J. T., Thompson, J. F., and Jarvis, N. (1980b). The mRNAs that code for soybean seed proteins. In: Genome Organisation and Expression in Plants. Leaver C. J. ed. Plenum: New York. pp.273-281.

Berdahl, J. D. and Bhatty, R. S. 1977. Protein fractions and lysine distribution in barley lines derived from hiproly. Can. J. Pl. Sci. 57, 1135-1139.

Bietz, J. A., Huebner, F. R. and Rothfus, J. A. (1970). Chromatographic comparison of peptic digests of individual gliadin proteins. Cereal Chem. 47, 393-404.

Bietz, J. A., Shepherd, K. W., and Wall, J. S. (1975) Single kernel analysis of glutenin: use in wheat genetics and breeding. Cereal Chem. 52, 513-532.

Bietz, J. A., Huebner, F. R., Sanderson, J. E. and Wall, J. S. (1977). Wheat gliadin homology revealed through N-terminal amino acid sequence analysis. Cereal Chem. 54, 1070-1083.

Bietz, J. A., Paulis, J. W. and Wall, J. S. (1979). Zein subunit homology revealed through amino-terminal sequence homology. Cereal Chem. 56, 327-332.

Bietz, J. A. (1980). Note on the amino-terminal amino acid sequence of hordein. Cereal Chem. 57. (In press).

Blagrove, R. J., Gillespie, J. M., and Randall, P. J. (1976). Effect of sulphur supply on the seed globulin composition of Lupinus angustifolius. Aust. J. Plant Physiol. 3, 173.

Blobel, G. and Dobberstein, B. (1975a). Transfer of proteins across membranes. J. Cell Biol. 67, 835-851.

Blobel, G. and Dobberstein, B. (1975b). Transfer of proteins across membranes II. J. Cell Biol. 67, 852-862.

Blobel, G., Walter, P., Chang, C. N., Goldman, B., Ericson, A. H., and Lingappa, V. R. (1979). Translocation of proteins across membranes: the signal hypothesis and beyond. In: Symposium of the Society for Experimental Biology 33. Hopkins, C. R. and Duncan, C. J. eds. Cambridge University Press. pp.9-36.

Blobel, G. (1980). Protein topogenesis. Proc. Natl. Acad. Sci. USA, 77, 1496-1500.

Bollini, R. and Chrispeels, M. J. (1979) The rough endoplasmic reticulum is the site of reserve-protein synthesis in developing Phaseolus vulgaris cotyledons. Planta, 487-501.

Boulter, D. (1979). Structure and biosynthesis of legume storage proteins. In Seed Protein Improvement in Cereals and Grain Legumes, Vol. I. IEAE; Vienna. pp. 125-134.

Boulter, D. (1980). Outogeny and development of biochemical and nutritional attributes in legume seeds. In: Advances in Legume Science, Summerfield R. and Bunting A. C. eds. pp. 127-134.

Brandt, A. (1979). Cloning of double stranded DNA coding for hordein polypeptides. Carlsberg Res. Commun. 44, 255-267.

Brandt, A. and Ingversen, J. (1976) Cloning of double stranded DNA coding for hordein polypeptides. Carlsberg Res. Commun. 41, 312-320.

Brandt, A. and Ingversen, J. (1978) In vitro synthesis of barley endosperm proteins on wild type and mutant templates. Carlsberg Res. Commun. 4, 451-469.

Bright, S. W. J., Lea, P. J., and Miflin, B. J. (1980). The regulation of methionine biosynthesis and metabolism in plants and bacteria. Ciba Foundation Symposium No. 72, Sulphur in Biology (Elliott, K. and Whelan J. eds.) Excerpta Medica:Amsterdam, pp.101-114.

Bright, S. W. J., Wood, E. A., and Miflin, B. J. (1978). The effect of aspartate-derived amino acids (lysine, threonine, methionine) on the growth of excised embryos of wheat and barley. Planta, 139, 113-117.

Brown, J. S. W., Kemble, R. J., Law, C. N., and Flavell, R. B. (1980). Control of endosperm proteins in Triticum aestivum L. (var. Chinese Spring) and Aegilops umbellata Zhuk. by homeologous group I chromosomes. Genetics (in press).

Burr, B and Burr, F. A. (1976). Zein synthesis in maize endosperm by polyribosomes attached to protein bodies. Proc. Nat. Acad. Sci. USA 73, 515-519.

Burr, B. (1979). Identification of zein structural genes in the maize genome. In: Seed Protein Improvement in Cereals and Grain Legumes, Vol. I. IAEA: Vienna. pp.175-178.

Burr, B., Burr, F. A., Rubenstein. I., and Simon, M. N. (1978). Purification and translation of zein messenger RNA from maize endosperm protein bodies. Proc. Nat. Acd. Sci. USA, 75, 696-700.

Burr, F. A. (1979). Zein synthesis and processing on zein protein body membranes. In: Seed Protein Improvement in Cereals and Grain Legumes Vol. I. IAEA: Vienna. pp.159-162.

Burr, F. A. and Burr, B. (1980). The cloning of zein sequences and an approach to zein genetics. In: Genome Organisation and Expression in Plants Leaver C. J. ed. Plenum: New York. pp.227-231.

Bushuk, W., Khan, K., and McMaster, G. (1980). Functional glutenin complex of covalently and non-covalently linked compounds. Annl. Instit. Natl. Rech. Agron. (in press).

Byers, M., Kirkman, M. A. and Miflin, B. J. (1977). Factors affecting the quality and yield of seed protein. In: Plant Proteins, G. Norton ed. Butterworths: London, pp. 227-243.

Cameron-Mills, V. and Ingversen, J. (1978). In vitro synthesis and transport of barley endoplasmic reticulum proteins: reconstitution of functional rough microsomes from polyribosomes and stripped microsomes. Carlsberg Res. Commun. 43, 471-489.

Cameron-Mills, V., Ingversen, J., and Brandt, A. (1978) Transfer of in vitro synthesized barley endosperm proteins into the lumen of the endoplasmic reticulum. Carlsberg Res. Commun. 43, 91-102.

Casey, R. (1979a). Immunoaffinity chromatography as a means of purifying legumin from Piseum (pea) seeds. Biochem. J. 177, 509-520.

Casey, R. (1979b) Genetic variability in the structure of the α-subunits of legumin from Pisum - a two-dimensional gel electrophoresis study. Heredity 43, 265-272.

Catsimpoolas, N. (1969). Isolation of glycinin sub-units by isoelectric focusing in urea-mercaptoethanol. FEBS Letts. 4, 259-261.

Catsimpoolas, N. and Wang, J. (1971). Analytical scanning isoelectric focussing 5. Separation of glycinin subunits in urea-dithiothreitol media. Anal. Biochem. 44, 436-444.

Charbonnier, L. (1973). Etudes des protéines alcoolo-solubles de la farine de blé. Biochimie 55, 1217-1225.

Croy, R. R. D., Gatehouse, J. A., Evans, I. M., and Boulter, D. (1980a). Characterisation of the storage protein subunits synthesized in vitro by polyribosomes and RNA from developing pea (Pisum sativum L.). I. Legumin. Planta 148, 49-56.

Croy, R. R. D., Gatehouse, J. A., Evans, I. M., and Boulter, D. (1980b). Characterization of the storage protein subunits synthesized in vitro by polyribosomes and RNA from developing pea (Pisum sativum L.) II. Vicilin. Planta 148, 57-63.

Dalby, A. and Tsai, C. Y. (1975). Comparisons of lysine and zein and non-zein protein contents in immature and mature maize endosperm mutants. Crop Sci. 15, 513-515.

Danielsson, C. E. (1949). Seed globulins of the gramineae and leguminosae. Biochem. J. 44, 387-400.

Davey, R. A. and Dudman, W. F. (1979). The carbohydrate content of storage glycoprotein from seeds of Pisum sativum: characterization and distribution on component polypeptides. Aust. J. of Plant Physiol. 6, 435-437.

Davies, D. R. (1976). Variation in the storage proteins of peas in relation to S amino acid content. Euphytica.25(3) 714-724.

Davies, D. R. (1980). The r_a locus and legumin synthesis in Pisum sativum. Biochem. Genet. (in press).

Davies, H. M. and Delmer, D. P. (1979). Seed reserve protein glycosylation in an in vitro preparation from developing cotyledons of Phaseolus vulgaris. Planta 146, 513-520.

Derbyshire, E., Wright, D. J. and Boulter, D. (1976). Legumin-vicilin storage proteins of legume seeds. Phytochem. 15, 3-24.

DiFonzo, N. and Stanca, A. M. (1977). EMS-derived barley mutants with increased lysine content. Genet. Agr. 31, 401-409.

DiFonzo, N., Fornasari, E., Salamini, F., Soave, C. (1977). SDS-protein subunits in normal, opaque-2 and floury 2 endosperms. Maydica, 22, 77-88.

DiFonzo, N., Gentinetta, E., Salamini, F. and Soave, C. (1979). Action of the opaque-7 mutation on the accumulation of storage protein products in maize endosperm. Pl. Sci. Lett. 14, 345-354.

Dieckert, J. W., and Dieckert, M. C. (1976). The chemistry and cell biology of the vacuolar proteins of seeds. J. Food Sci. 41, 475-482.

Doekes, G. J. (1973). Inheritance of gliadin composition in bread wheat, Triticum aestivum L. Euphytica, 22, 28-33.

Doll, H. (1980) A nearly non-functional mutant allele of the storage protein locus Hor 2 in barley. Theor. Appl. Genet. (in press).

Doll, H. and Brown, A. D. H. (1979). Hordein variation in wild (Hordeum spontaneum) and cultivated (H. vulgare) barley. Can. J. Genet. Cytol. 21, 391-404.

Doll, H., Koie, B. and Eggum, B. O. (1974). Induced high-lysine mutants in barley. Radiation Bot. 14, 73-80.

Domoney, C., Davies, D. R., and Casey, R. (1980). The initiation of legumin synthesis in immature embryos of Pisum sativum L. grown in vivo and in vitro. Planta (in press)

Dudman, W. F. and Millerd, A. (1975). Immunochemical behaviour of legumin and vicilin from Vicia faba survey of related proteins in the leguminosae sub family Fabrideae. Biochem. Syst. Ecol. 3, 25-33.

Ericson, M. C. and Chrispeels, M. J. (1973). Isolation and characterization of glucosamine-containing storage glyco-proteins from the cotyledons of Phaseolus aureus. Plant Physiol. 52, 98-104.

Ericson, M. C. and Chrispeels, M. J. (1976). The carbohydrate moiety of mung bean vicilin. Aust. J. Plant Physiol. 3, 763-769.

FAO (1970). Amino Acid Content of Foods and Biological Data on Proteins. FAO:Rome.

FAO (1979). Production Yearbook (1978), Vol. 32. FAO: Rome.

Fox, J. E., Pratt, H. M., Shewry, P. R., and Miflin, B. J. (1977). The in vitro synthesis of hordeins with polysomes from normal and high lysine varieties of barley. In: Nucleic Acids and Protein Synthesis in Plants. CNRS:Paris, pp.501-509.

Gatehouse, J. A. and Croy, R. R. D. and Boulter, D. (1980). Isoelectrifocussing properties and carbohydrate content of pea (Pisum sativum) legumin. Biochem. J. 185, 497-503.

Gayler, K. R. and Sykes, G. E. (1980). β-conglycinin in developing soybean seeds. Plant Physiol. (in press).

Gianazza, E., Righetti, P. G., Pioli, F., Galante, E. and Soave, C. (1976). Size and charge heterogeneity of zein in normal and opaque 2 endosperms. Maydica 21, 1-17.

Gilroy, J., Wright, D. J., and Boulter, D. (1979). Homology of basic subunits of legumin from Glycine max and Vicia faba. Phytochemistry 18, 315-326.

Guiragossian, V., Chibber, B. A. K., van Scoyoc, S., Jambunathan, R., Mertz, E. T. and Axtell, J. D., (1978). Characteristics of proteins from normal, high lysine and high tannin sorghums. J. Agric. Food Chem., 26, 219-223.

Hall, T. C., Ma, Y., Buchbinder, B. U., Pyne, J. W., Sun, S. M. and Bliss, F. A. (1978). Messenger RNA for G1 protein of French bean seeds: Cell-free translation and product characterization. Proc. Natl. Acad. Sci. 75, 3196-3200.

Hall, T. C., McLeester, R. C. and Bliss, F. A. (1977). Plant Physiol. 59, 1122-1124.

Hall, T. C., Sun, S. M., Buchbinder, B. U., Pyne, J. W., Bliss, F. A. and Kemp, J. D. (1980). Bean seed globulin in mRNA: Translation, characerization and its use as a proble towards genetic engineering of crop plants. In: Genome Organisation and Expression in Plants. Leaver, C. J. ed. Plenum: New York. pp.259-272.

Higgins, T. J. and Spencer, D. (1977). Cell free synthesis of pea seed proteins. Plant Physiol. 60, 655-661.

Higgins, T. J. V. and Spencer, D. (1980). Biosynthesis of pea seed proteins: evidence for precursor forms from in vivo and in vitro studies. In: Genome Organisation and Expression in Plants. (1980). Leaver, C. J. ed. Plenum:New York. pp.245-258.

Holder, A. A. and Ingversen, J. (1978). Peptide mapping of major components of in vitro synthesized barley hordeins. Evidence of structural homology. Carlsberg Res. Commun. 43, 177-184.

Hynes, M. J. (1968). Genetically controlled electrophoretic variants of a storage protein in Pisum sativum. Aust. J. Biol. Sci. 21, 827-829.

Ingversen, J. (1975). Structure and composition of protein bodies from wild type and high-lysine barley endosperm. Hereditas 81, 69-76.

Ingversen, J. and Koie, B. (1973). Lysine rich proteins in high-lysine H. vulgare grain. Phytochemistry, 12, 1107-1111.

Jensen, C. J. (1976). Barley monoploids and doubled monoploids: techniques and experience. In Barley Genetics III. H. Gaul ed. pp.316-345. Karl Thiemig, Munich.

Jensen, J. (1979a). Location of a high-lysine gene and the DDT-resistance gene on barley chromosome 5. Euphytica 28, 47.

Jensen, J. (1979b). Chromosomal location of one dominant and four recessive high-lysine mutants. In: Seed Protein Improvement in Cereals and Grain Legumes. IAEA:Vienna. pp.89-96.

Jensen, J., Jorgensen, J. H., Jensen, H. P., Giese, H. and Doll, H. (1980). Linkage of the hordein loci Hor 1 and Hor 2 with the powdery mildew resistance loci Ml-k and Ml-a on barley chromosome 5. Theor. Appl. Genet. (In press).

Johnson, V. A., Mattern, P. J., and Kuhr, S. L. (1979). Genetic improvement of wheat protein. In: Seed Protein Improvement in Cereals and Grain Legumes, Vol. II. IAEA: Vienna. pp.165-179.

Jones, R. A. (1978). Biochem. Genet. 16, 27-38.

Karlsson, K. E. (1976). In "Barley Genetics III" (H. Gaul, ed.) Garching, pp.536-541.

Karlsson, K. E. (1977). Barley Genetics Newsletter 7, 40-43.

Kasarda, D. D., Da Roza, D. A. and Ohms, J. I. (1974). N-terminal sequence of α_2-gliadin. Biochim. Biophys. Acta, 351, 290-294.

Kasarda, D. D., Bernardin, J. E. and Qualset, C. O. (1976). Relationship of gliadin protein components to chromosomes in hexaploid wheats. Proc. Natl. Acad. Sci. USA, 73, 3646-50.

Kasarda, D. D., Bernardin, J., and Nimmo, C. C. (1977). Wheat proteins. In: Advances in Cereal Science and Technology, Y. Pomeranz, ed. American Assoc. Cereal Chem:Minnesota, U.S.A., pp.158-

Kasarda, D. (1980). Structure and properties of gliadins. Ann. Instit. Natl. Rech. Agron. (in press).

Khoo, U. and Wolf, M. J. (1970). Origin and development of protein granules in maize endosperm. Amer. J. Bot. 57, 1042-1050.

Kitamura, K. and Shibasaki, K. (1975). Homology between the acidic subunits of soybean 11S globulin. Agr. Biol. Chem. 39, 1509-1510.

Kitamura, K. and Shibasaki, K. (1977). Homology among the acidic subunits of soybean 11S globulin. Agr. Biol. Chem. 41, 351-357.

Koie, B. and Doll, H. (1979). Protein and carbohydrate components in the Riso high-lysine barley mutants. In: Seed Protein Improvement in Cereals and Grain Legumes. IAEA:Vienna. pp.205-215

Koie, B. and Kreis, M. (1978). Hordein and starch synthesis in developing high lysine and normal barley seeds at different N-fertilizer levels. In: Carbohydrate and Protein Synthesis (B. J. Miflin and M. Zoschke eds.) EEC, Luxembourg, pp.137-154.

Koie, B., Doll, H. and Kreis, M. (1976). Evaluation of a high-lysine barley gene by chromosome double monoploids. Genetika 8, 177-182.

Koshiyama, I.and Fukushima (1976). A note on carbohydrates in the US globulin of soybean seeds. Cereal Chem. 53, 768-769.

Kreis, M. (1978). Starch and free sugars during kernel development of Bomi barley and its high-lysine mutant 1508. In: Seed Protein Improvement by Nuclear Techniques. IAEA, Vienna, pp.115-120.

Lane, C. D., Colman, A., Mohun, T., Morser, J., Champion, J., Kourides, I., Craig, R., Higgins, S., James, T. C., Applebaum, S. W., Ohlsson, R. I., Paucha, E., Houghton, M., Matthews, J. and Miflin, B. J. (1980). The Xenopus oocyte as a surrogate secretory system: the specificity of protein export. Eur. J. Biochem. (in press).

Larkins, B. A. and Dalby, A. (1975) In vitro synthesis of zein like protein by maize polyribosomes. Biochim. Biophys. Res. Commun. 66, 3

Larkins, B. A. and Hurkman, W. J. (1978). Synthesis and deposition of zein in protein bodies of maize endosperm. Plant Physiol. 62, 256-263.

Larkins, B. A., Bracker, C. E., and Tsai, C. Y. (1976a) Storage protein synthesis in maize - isolation of zein synthesising polyribosomes. Plant Physiol. 57, 740-745.

Larkins, B. A., Jones, R. A., and Tsai, C. Y. (1976b). Isolation and in vitro translation of zein messenger ribonucelic acid. Biochemistry 15, 5506-5511.

Larkins, B. A., Pedersen, K., Handa, A. K., Hurkman, W. J. and Smith, I. D. (1979). Synthesis and processing of maize storage proteins in Xenopus laevis oocytes. Proc. Natl. Acad. Sci. USA. 76, 6448-6452.

Larkins, B. A., Pedersen, K., Hurkman, W. J., Handa, A. K., Mason, A. C., Tsai, C. Y., and Hermondson, M. A. (1980). Maize storage proteins: characterization and biosynthesis. In: Genome Organisation and Expression in Plants. Leaver C. J. ed. Plenum: New York. pp.203-217.

Law, C. N. Young, C. F., Brown, J. W. S., Snape, J. W., and Worland, A. J. (1978). The study of grain protein control in wheat using whole chromosome substitution lines. In: Seed Protein Improvement by Nuclear Techniques. IAEA: Vienna. pp.483-502.

Lee, K. H., Jones, R. A., Dalby, A. and Tsai, C. Y. (1976). Genetic regulation of storage protein content in maize endosperm. Biochem. Genet. 14, 641.

Lingappa, V. R., Lingappa, J. R., and Blobel, G. (1979). Chicken ovalbumin contains an internal signal sequence. Nature 281, 117-121.

Ma, Y. and Nelson, O. E. (1975). Amino acid composition and storage proteins in two new high lysine mutants in maize. Cereal Chem. 52, 412-419.

Manteuffel, R. and Scholz, G. (1975). Studies on seed globulins from legumes V. Immunoelectrophoretic control of vicilin purification by gel filtration. Biochem. Physiol. Pflanzen 168, 277-285.

Marshall, R. D. (1972). Glycoproteins. Ann. Rev. Biochem. 41, 673-702.

Matile Ph. (1976). Vacuoles. In: Plant Biochemistry, 3rd ed. Bonner, J. and Varner, J. E. eds. Academic Press:New York, pp.189-224.

Matthews, J. A. and Miflin, B. J. (1980). In vitro synthesis of barley storage proteins. Planta (in press).

McLeester, R. C., Hall, T. C., Sun, S. M., and Bliss, F. A. (1973). Comparison of globulin proteins from Phaseolus vulgaris with those from Vicia faba. Phytochemistry 12, 85-93.

McWhirter, K. S. (1971). A floury endosperm, high lysine locus on chromosome 10. Maize Genet. Coop. Newsletter, 45, 184.

Mecham, D. K., Kasarda, D. D. and Qualset, C. O. (1978). Genetic aspects of wheat gliadin protein. Biochem. Genet. 16, 831-853.

Mertz, E. T. (1976). Case histories of existing models. In: Genetic Improvement of Seed Proteins, National Academy of Sciences. Washington D.C. pp.57-70.

Mertz, E. T., Bates, L. S. and Nelson, O. E. (1964). Mutant gene that changes protein composition and increases lysine content of maize endosperm. Science, 145, 279-280.

Miflin, B. J. (1976). Metabolic control of biosynthesis of nutritionally essential amino acids. In: Genetic Improvement of Seed Proteins. National Academy of Sciences USA: Washington. pp.135-155.

Miflin, B. J. and Shewry, P. R. (1977). eds. Techniques for the Separation of Barley and Maize Proteins. Commission of the European Communities: Luxembourg, 114 pp.

Miflin, B. J. and Shewry, P. R. (1979a). The synthesis of proteins in normal and high lysine barley seeds. In: Recent Advances in the Biochemistry of Cereals. Laidman, D. and Wyn Jones, R.G. eds. Academic Press, London pp.239-273.

Miflin, B. J. and Shewry, P. R. (1979b). The biology and biochemistry of cereal seed prolamins. In Seed Improvement in Cereals and Grain Legumes, Vol. I. IAEA:Vienna, pp.137-158.

Miflin, B. J., Burgess, S. R. and Shewry, P. R. (1980a). The development of protein bodies in the storage tissues of seeds: Subcellular separations of homogenates of barley, maize and wheat endosperms and of pea cotyledons. J. Exp. Bot. (in press).

Miflin, B. J., Byers, M., Field, J. M., Faulks, A. J., and Forde, J. (1980b). The isolation and characterization of proteins extracted from whole milled seed, gluten and developing protein bodies of wheat. Annl. Instit. Natl. Rech. Agron. (in press).

Miflin, B. J., Matthews, J. A., Burgess, S. R., Faulks, A. J. and Shewry, P. R., (1980c). The synthesis of barley storage proteins. In: Genome Organisation and Expression in Plants. (ed. Leaver C. J.) pp.233-243.

Millerd, A., Thomson, J. A., Schroeder, H. E. (1978). Cotyledonary storage proteins in Pisum sativum III. Patterns of accumulation during development. Aust. J. Plant Physiol. 5, 519-534.

Millerd, A., Thomson, J. A. and Randall, P. J. (1979). Heterogeneity of sulphur content in the storage proteins of pea cotyledons. Planta, 146, 463-466.

Misra, P. S., Mertz, E. T. and Glover, D. V. (1975). Studies on corn proteins VI. Endosperm protein changes in single and double endosperm mutants of maize. Cereal Chem. 52, 161-166.

Moreira, M. A., Hermodson, M. A., Larkins, B. A., and Nielsen, N. C. (1979). Partial characterization of the acidic and basic polypeptides of glycinin. J. Biol. Chem. 254, 9921-9926.

Morton, R. K., Palk, B. A., and Raison, I. K. (1964). Intracellular components associated with protein synthesis in developing wheat endosperm. Biochem. J. 91, 522-528.

Munck, L.(1972). Improvement of nutritional value in cereals. Hereditas 72, 1-128.

Munck, L. and Von Wettstein, D. (1976). Effects of genes that change the amino acid composition of barley endosperm. In: Genetic Improvement of Seed Protein. National Acad. Sciences, Washington, D.C., pp. 71-82.

Munck, L., Karlsson, K. E., Hagberg, A. C., Eggum, B. O. (1970). Gene for improved nutritional value in barley seed protein. Science 168, 985-987.

Murray, D. R. (1979). A storage role for albumin in pea cotyledons. Plant Cell and Environment 2, 221-226.

Nagahasi, J. and Beevers L. (1978). Subcellular localization of glycoly transferases involved in glycoprotein biosynthesis in the cotyledons of Pisum sativum L. Plant Physiol. 61, 451-459.

Nelson, O. E. (1979). Inheritance of amino acid content in cereals. In: Seed Protein Improvement in Cereals and Grain Legumes. IAEA:Vienna. pp.79-87.

Nelson, O. E. (1980). Genetic control of polysaccharide and storage protein synthesis in the endosperms of barley, maize and sorghum. In: Advances in Cereal Science and Technology Vol. 3. Y. Pomeranz, ed. American Association of Cereal Chemists, St. Paul, Minn. pp. (In press)

Nelson, O. E., Mertz, E. T. and Bates, L. S. (1965). Science, 150, 1469-70.

Neumann, D. and Weber, E. (1978). Formation of protein bodies in ripening seeds of Vicia faba L. Biochem. Physiol. Pflanzen. 173, 167-180.

Okubo, K., Asano, M., Kimura, Y., and Shibasaki, K. (1969). On basic subunits dissociated from 11S component of soybean proteins with urea. Agric. Biol. Chem. 33, 463-465.

Ooynen, A. van., Mantei, N., Berg, J. van der, Beggs, J. D., Boll, W., Weaver, R. F., and Weissman, C. (1979). The expression of a cloned rabbit chromosomal β-globin gene in mouse L cells and yeast. In: Recombinant DNA and Genetic Experimentation. Morgan, J. and Whelan, W. J. eds. Pergammon:Oxford. pp.141-146.

Oram, R. N., Doll, H. and Koie, B. (1975). Genetics of two storage protein variants in barley. Hereditas, 80, 53-58.

Osborne, T. B. (1924). Vegetable Proteins. Longmans Green:London pp.154.

Osborne, T. B. and Campbell, G. R. (1898). Proteids of the pea. J. Am. Chem. Soc. 29, 348-362.

Park, W. D., Lewis, E. D. and Rubinstein, I. (1980). Plant Physiol. 65, 98-106.

Paulis, J. W. and Wall, J. S. (1977). Comparison of the protein composition of selected corms and their wild relatives, Teosinto and Tripsacum. J. Agric. Food Chem. 25, 265-270.

Paulis, J. W. and Wall, J. S. (1979). Distribution and electrophoretic properties of alcohol-soluble proteins in normal and high-lysine sorghum. Cereal Chem. 56, 20-24.

Payne, P. I., Law, C. N., and Mudd, E. E. (1980). Control by homeologous group chromosomes of the high-molecular weight subunits of glutenin, a major protein of wheat endosperm. Theor. Appl. Genet. 58, in press.

Pedersen, K., Bloon, K. S., Anderson, J. N., Glover, D. V., and Larkins, B. A. (1980). Analysis of the complexity and frequency of zein gene in the maize genome. Biochemistry, 19, 1644-50.

Pernollet, J-C. (1978). Protein bodies of seeds: Ultrastructure biochemistry, biosynthesis and degradation. Phytochem. 17, 1473-1480.

Persson, G. and Karlsson, K. (1977). Progress in breeding for improved nutritive value in barley. Cereal Res. Commun. 5, 169-179.

Peterson, D. M. and Smith, D. (1976). Changes in nitrogen and carbohydrate fractions in developing oat groats. Crop Science 16, 67-71.

Püchel, M., Müntz, K., Parthier, B., Aurich, O., Bassüner, R., Manteufel, R., Schmidt, P. (1979). RNA metabolism and membrane-bound polysomes in relation to globulin biosynthesis in cotyledons of developing field beans (Vicia faba L.). Eur. J. Biochem. 96, 321-329.

Pusztai, A. and Watt, W. B. (1970). Glytcoprotein II: The isolation and characterization of a major antigenic and non-haemaglutinating glycoprotein from Phaseolus vulgaris. Biochim. Biophys. Acta 207, 413-431.

Righetti, P. G., Gianazza, E., Viotti, A., and Soave, C. (1977). Heterogeneity of storage proteins in maize. Planta, 136, 115-.

Romero, J., Sun, S-M. M. McLeester, R. C., Bliss, F. A. and Hall, T. C. (1975). Heritable variation in a polypeptide subunit of the major storage protein of the bean, Phaseolus vulgaris L. Plant Physiol. 56, 776-779.

Rybalka, A. (1980). Genetic basis of gliadin and hordein nomenclature. Annl. Instit. Natl. Rech. Agron. (in press).
Salamini, F., DiFonzo, N., Gentinetta, E., and Soave, C. (1979). A dominant mutation interfering with protein accumulation in maize seed. In: Seed Protein Improvement in Cereals and Grain Legumes, Vol. I, IAEA: Vienna. pp. 97-108.
Schmitt, J. M. (1979). Purification of hordein polypeptides by column chromatography using volatile solvents. Carlsberg Res. Commun. 44, 431-438.
Schmitt, J. M. and Svendsen, I. (1980). Amino acid sequences of hordein polypeptides. Carlsberg Res. Commun. 45, 143-148.
Shepherd, K. W. (1968). Chromosomal control of endosperm protein in wheat and rye. In: Proc. 3rd Int. Wheat Genetics Symp. Finlay, K. W. C., Shepherd, K. W., eds. Butterworths: London. pp.86-96.
Shepherd, K. W. (1973). Homeology of wheat and alien chromosomes controlling endosperm protein phenotypes. In: Proc. 4th Int. Wheat Genet. Symp. (Sears, E. R. and Sears, L. M. S. eds.) Agric. Exp. Station, University of Missouri, Columbia, Mo. pp. 745-760.
Shewry, P. R., Ellis, J. R. S., Pratt, H. M., and Miflin, B. J. (1978a). A comparison of methods for the extraction and separation of hordein fractions from 29 barley varieties. J. Sci. Food Agric. 29, 433-441.
Shewry, P. R., Hill, J. M., Pratt, H. M., Leggatt, M. M., and Miflin, B. J. (1978b). An evaluation of techniques for the extraction and separation of hordein and glutelin from barley seed and a comparison of the protein composition of Bomi and Riso 1508. J. Exp. Bot. 29, 677-692.
Shewry, P. R., Pratt, H. M., Finch, R. A. and Miflin, B. J. (1978c). Genetic analysis of hordein polypeptides from single seeds of barley. Heredity, 40, 463-466.
Shewry, P. R., Pratt, H. M., Leggatt, M. M., and Miflin, B. J. (1979). Protein metabolism in developing endosperms of high-lysine and normal barley. Cereal Chem. 56, 110-117.
Shewry, P. R., Autran, J. C., Nimmo, C. C., Lew, E. J-L. and Kasarda, D. D. (1980a). N-terminal amino acid sequence homology of storage protein components (prolamins) from barley (Hordeum vulgare L.) and a diploid wheat (Triticum monococcum L.). Nature (in press).
Shewry, P. R., Faulks, A. J. and Miflin, B. J. (1980b). The effect of high-lysine mutations on the protein fractions of barley grain. Biochem. Genet., 18, 133-
Shewry, P. R., Faulks, A. J., Pickering, R. A., Jones, I. T., Finch, R. A. and Miflin, B. J., (1980c). The genetic analysis of barley storage proteins. Heredity, 44, 383-389.
Shewry, P. R., Field, J. M., Kirkman, M. A., Faulks, A. J. and Miflin, B. J. (1980d). The extraction, solubility and characterization of two groups of barley storage polypeptides. J. Exp. Bot. 31, 393-407.
Shewry, P. R., March, J. F. and Miflin, B. J. (1980e). N-terminal amino acids sequence of C hordein. Phytochem. (in press).
Singh, R. and Axtell, J. D. (1973). High lysine mutant gene (hl) that improves protein quality and biological value of grain sorghum. Crop Sci. 13, 535-539.
Smith, D. B. and Flavell, R. B. (1974). The relatedness and evolution of repeated nucleotide sequences in the genomes of some Graminae species. Biochem. Genet. 12, 243-256.
Soave, C., Righetti, P. G., Lorenzoni, C., Gentinetta, E., Salamini, F. (1976). Expressivity of the opaque-2 gene at the level of zein molecular components. Maydica, 21, 61-75.

Soave, C., Suman, N., Viotti, A. and Salamini, F. (1978). Linkage relationships between regulatory and structural gene loci involved in zein synthesis in maize. Theor. Appl. Genet. 52, 263-267.

Soave, C., Viotti, A., DiFonzo, N., and Salamini, F. (1979). Maize prolamin: synthesis and genetic regulation. In: Seed Protein Improvement in Cereals and Grain Legumes, Vol. I. IAEA: Vienna. pp.165-173.

Soave, C., Viotti, A., DiFonzo, N., Salamini, F. (1980). Recent evidence concerning the genetic regulation of maize synthesis. In: Genome Organisation and Expression in Plants, Leaver C. J. ed. Plenum:New York. pp.219-226.

Stafford, A. and Davies, D. R. (1979). The culture of immature pea embryos. Ann. Bot. 44, 315-321.

Sun, S. M. and Hall, T. C. (1975). Solubility characteristics of globulins from Phaseolus seeds in regard to their isolation and characterization. J. Agric. Food Chem. 23, 2.

Sun, S. M., Buchbinder, B. U., and Hall, T. C. (1975). Cell free synthesis of the major storage protein in bean Phaseolus vulgare. Plant Physiol. 56, 780-785.

Sun, S. M., Mutschler, M. A., Bliss, F. A., and Hall, T. C. (1978). Protein synthesis and accumulation in bean cotyledons during growth. Plant Physiol. 61, 918-923.

Sykes, G. E. and Gayler, K. R. (1980). Detection and characterization of a new beta-conglycinin from soybean seeds. Manuscript under review. Arch. Biochem. Biophys.

Thanh, V. H. and Shibasaki, K. (1976a). Major proteins of soybean seeds. A straightforward fractionation and their characterization. J. Agric. Fd. Chem. 24, 1117-1121.

Thanh, V. H. and Shibasaki, K. (1976b). Heterogeneity of Beta-conglycinin. Biochim. Biophys. Acta, 439, 326-338.

Thanh, V. H. and Shibasaki, K. (1977). Beta-conglycinin from soybean proteins. Biochim. Biophys. Acta 490, 370-384.

Thanh, V. H. and Shibasaki, K. (1978a). Major proteins of soybean seeds. Subunit structure of β-conglycinin. J. Agric. Fd. Chem. 26, 692-695.

Thanh, V. H. and Shibasaki, K. (1978b). Major proteins of soyabean seeds. Reconstitution of β-conglycinin from its subunits. J. Agric. Fd. Chem. 26 695-698.

Thomson, J. A. (1975). Major patterns of gene activity during development in holometabolous insects. Adv. Insect Physiol. 11, 321.

Thomson, J. A. and Doll, H. (1979). Genetics and evolution of seed storage proteins. In: Seed Protein Improvement in Cereals and Grain Legumes Vol. I. IAEA: Vienna. pp. 109-123.

Thomson, J. A. and Schroeder, H. E. (1978). Cotyledonary storage proteins in Pisum sativum II. Hereditary variation in components of the legumin and vicilin fractions. Aust. J. Plant Physiol. 5, 263-279.

Thomson, J. A., Schroeder, H. E. and Dudman, W. F. (1978). Storage proteins in Pisum sativum I. Molecular heterogeneity. Aust. J. Plant Physiol. 5, 263-279.

Trowbridge, I. S. (1974). Isolation and chemical characterization of a mitogenic lectin from Pisum sativum. J. Biol. Chem. 249, 6004-6012.

Tsai, C. Y., Larkins, B. A. and Glover, D. V. (1978). Interaction of the opaque-2 gene with starch-forming mutant genes on the synthesis of zein in maize endosperm. Biochem. Genet. 16, 883-896.

Valentini, G., Soave, C. and Ottaviano, E. (1979). Chromosomal location of zein genes in maize. Heredity, 42, 33-40.

Vasal, S. K. (1975). In High protein quality maize. Dowden, Hutchinson and Ross, Stroudsberg, Penn. pp. 197-216.

Vasil, V. and Vasil, I. K. (1980). Isolation and culture of cereal protoplasts II. Embryogenesis and plantlet formation from protoplasts of Pennisetum americanum. Theor. Appl. Genet. 56 (3) 97-100.

Villareal, R. M. and Juliana, B. J. (1978) Properties of glutelin from mature and developing rice grain. Phytochem. 17, 177-182.

Viotti, A., Pogna, N. E., Balducci, C., Durante, M. (1980). Chromosomal localization of zein genes by in situ hybridization in Zea mays. Mol. Gen. Genet. (in press).

Viotti, A., Sala, E., Alberi, P., and Soave, C. (1978). Heterogeneity of zeins synthesized in vitro. Pl. Sci. Letts. 13, 365-375.

Viotti, A., Sala, E., Marotta, R., Alberi, P., Balducci, C., Soave, C. 1979. Genes and mRNAs coding for zein polypeptides in Zea mays. Eur. J. Biochem. 102, 211-222.

Wall, J. S. (1979). The role of wheat proteins in determining baking quality. In: Recent Advances in the Biochemistry of Cereals. Laidman, D. L. and Wyn Jones, R. G. eds. Academic Press: London, pp.275-312.

Weinand, U. and Feix, G. (1978). Electrophoretic fractionation and translation in vitro of poly (rA)-containing RNA from maize endosperm. Eur. J. Biochem. 92, 605-611.

Weinand, U., Brüschke, C., and Feix, G. (1979). Cloning of double stranded DNAs derived from polysomal mRNA of maize endosperm: isolation and characterization of zein clones. Nuc. Acid. Res. 6, 2707.

Wilden, M. van der, Herman, E. M., and Chrispeels, M. J. (1979). Protein bodies contain a variety of acid hydrolases and constitute the general lytic compartment in the cotyledons of the mung bean. Plant Physiol. 63(5), S29.

Wilson, C. M. (1966). Bacteria, antibiotics and amino acids incorporation into maize endosperm protein bodies. Plant Physiol. 41, 325-327.

Wilson, C. M., Shewry, P. R., and Miflin, B. J. (1980). Maize endosperm proteins compared by sodium dodecyl sulphate gel electrophoresis and isoelectric focussing. Cereal Chem. (submitted).

Wright, D. J. and Boulter, D. (1974). Purification and subunit structure of legumin of Vicia faba (Broad bean). Biochem. J. 141, 413-418.

Wright, D. J. and Boulter, D. (1972). The characterization of vicilin during seed development in Vicia faba (L). Planta 105, 60-65.

Wrigley, C. W. and Shepherd, K. W. 1973. Electrofocusing of grain proteins from wheat genotypes. Ann. N.Y. Acad. Sci. 209, 154-162.

Yamauchi, F., Kawase, M., Kanbe, M., and Shibasaki (1975). Separation of the β-aspartamido CHO fraction from soyabean 7S proteins: Protein CHO linkage. Agric. Biol. Chem. 873-878.

Yamauchi, F., Thanh, V. H, Kawase, M., and Shibasaki, K. (1976). Separation of the glycopeptide from soyabean 7S proteins: their amino acid sequences. Agric. Biol. Chem. 40, 691-696.